PHYSICAL Geography LABORATORY MANUAL

exercises in atmospheric and earth surface processes

SIXTH EDITION
revised printing

David Shankman
Rhodes College

Carl A. Reese
University of Southern Mississippi

Cover image © Shutterstock.com

Kendall Hunt
publishing company

www.kendallhunt.com
Send all inquiries to:
4050 Westmark Drive
Dubuque, IA 52004-1840

Copyright © 1995, 1998, 2001 by David Shankman
Copyright © 2003, 2005, 2008 by Kendall Hunt Publishing Company
Revised Printing: 2013

ISBN 978-1-4652-2258-9

All rights reserved. No part of this publication may be reproduced,
stored in a retrieval system, or transmitted, in any form or by any means,
electronic, mechanical, photocopying, recording, or otherwise,
without the prior written permission of the copyright owner.

Printed in the United States of America
10 9 8 7 6 5 4

Contents

Preface v
Acknowledgments vii
About the Authors ix

Lab Exercises

1. Field Mapping   1
2. Latitude, Longitude, and Time   9
3. Earth-Sun Relations   27
4. Solar Radiation   39
5. Changing Earth-Sun Relations: Milankovitch Cycles   45
6. Atmospheric Pressure and Winds   53
7. Global Atmosphere and Ocean Circulation   63
8. Atmospheric Humidity   79
9. Adiabatic Processes and Precipitation   87
10. Air Masses, Cyclonic Storms, and Fronts   95
11. Tropical Cyclones and Hurricanes   109
12. Weather Analysis   123
13. Climate Classification   129
14. Topographic Map and Aerial Photograph Interpretation   149
15. Global Positioning System Applications   169
16. Trigonometry Applications in Geographic Field Work   179
17. Contour Mapping   189
18. Soil Properties   201
19. Soil Erosion   217
20. Runoff and Infiltration   231
21. Stream Discharge   237
22. Recurrence Intervals   249

23. Fluvial Geomorphology 263
24. Arid Landscapes 285
25. Glaciated Landscapes 301
26. Coastal Geomorphology 317

Appendices

Appendix A Trigonometry Tables 329

Appendix B Diameter and Corresponding Circumference and Area 331

Index 333

Preface

This manual is designed for a one or two semester laboratory course in physical geography or environmental science. It will complement most introductory physical geography textbooks. There are 26 laboratory exercises. The first thirteen exercises focus on earth-sun relationships and atmospheric processes. The last thirteen exercises deal with mapping, water resources, erosion processes, and landforms. The exercises are self-explanatory and include maps, aerial photographs, and worksheets. The exercises are designed, however, so the lab instructors can, if desired, provide their own maps and aerial photographs to show areas near your location.

This manual is an important alternative to those that are primarily workbooks intended to review classroom lectures. This manual is designed to give students a hands-on, field experience for a more complete understanding of the physical landscape. Many of the exercises in this manual require data collection from the field. Sample data sets are provided, however, if fieldwork is not practical. In other exercises, data are generated from interpretation of maps and aerial photographs. The data collection, mapping, and analysis required in this manual, encourages students to formulate questions and think critically about their observations, to recognize patterns in our physical environment, and better understand the processes that produce them.

The final four exercises include stereo aerial photographs and matching topographic maps. These exercises cover (1) fluvial geomorphology, (2) arid landscapes, (3) glaciated landscapes, and (4) coastal geomorphology. These photographs and maps and the accompanying problems provide reviews for classroom lectures in earth surface processes.

David Shankman
Visiting Professor in the Environmental Studies and Sciences Program, Rhodes College

Carl A. Reese
Professor of Geography, University of Southern Mississippi

Acknowledgments

The exercises in this book were tested in introductory physical geography laboratory sections at the University of Alabama. A special note of thanks is extended to the lab instructors, David Anderson, Justin Hart, Sara Hart, Maria Jose Garcia Quijano, Sara Mace, and Samuel Stutsman, whose suggestions and constructive criticisms greatly improved this manual.

Thomas Kallsen, Supervisor of the University of Alabama Map Library, provided time and helpful comments. We are indebted to Luoheng Han, W. Craig Remington, Qiaoli Liang, and Daniel P. Royall, who provided assistance in many ways. We are grateful for their help.

The figures were prepared by the University of Alabama Cartographic Research Lab.

About the Authors

David Shankman completed his doctoral study at the University of Colorado. He is currently Visiting Professor at Rhodes College. Previously he was Professor of Geography and Director of the Environmental Science Program at the University of Alabama. He teaches courses in biogeography, climatology, and geomorphology.

The author has active research programs dealing with river processes and bottomland hardwood ecosystems. He worked for many years in the southeastern U.S. Coastal Plain investigating flood control and modification of alluvial rivers, and the effects of river processes on floodplain forest vegetation. More recently, the author has been working in China as the principal investigator of a research project addressing flood control along the Yangtze River. He has published extensively in geography and interdisciplinary science journals.

Carl A. Reese completed his doctoral study at Louisiana State University. He is currently serving as a Professor of Geography at the University of Southern Mississippi, where he teaches courses in biogeography, climatology, paleoecology, statistics, and serves as the graduate coordinator.

Dr. Reese's primary research interest is the impact of climatic change on the vegetation of the High-Central Andes region of South America. His specialty involves the use of fossil pollen found in high-alpine ice caps and lakes to reconstruct paleovegetation. This research is currently being used to answer a wide variety of paleoecological questions centered on the timing, magnitude, and nature of climatic events from the late Pleistocene to the present. He is also actively involved in the new and exciting science of paleotempestology, or the reconstruction of pre-historic hurricane landfalls. He has worked all along the Gulf and Atlantic Coasts of North America and has recently extended this research into the Caribbean and Central America.

Name _____ Section _____

LAB EXERCISE

Field Mapping

1

The ability to determine distance, direction, and construct basic maps in the field is an important tool of geographers. Maps, signs, or other navigation tools are available in most areas. But in remote locations, where these tools are not available, you may need to know how to determine your position. The purpose of this exercise is to introduce skills necessary for mapping and for you to construct a map based on your observations and notes taken in the field.

Compass and Measuring Azimuth

Direction will be determined by using a compass. Compass directions are measured in degrees, from 0° to 360°. The four cardinal points of the compass are north (0°), east (90°), south (180°), and west (270°). The four cardinal points of a compass are shown in Figure 1.1. An **azimuth** is a compass direction based on a 360° circle. The needle of a compass points toward the magnetic north, which is not exactly the same as the north pole.

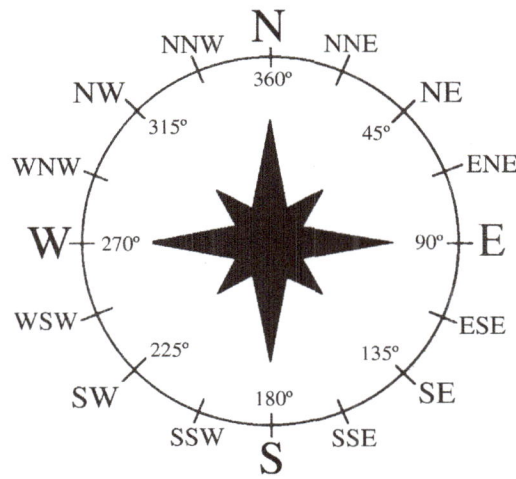

FIGURE 1.1 Four cardinal points of the compass.

1

To shoot an azimuth, hold the compass about waist level so that the direction arrow points directly away from you. Rotate the compass housing so that the needle aligns with orienting arrow (or 0° on the compass housing). The degree mark on the housing that is aligned with the direction arrow is your azimuth. Be sure to hold the compass level so that the needle can swing freely.

Two azimuth readings are usually taken to insure the accuracy. After sighting from point A to point B, read an azimuth in the opposite direction, from point B to point A. If the measurements are accurate, there will be a 180 degree difference.

Measure 3 azimuths based on sites given by the lab instructor.

Positions	Azimuth	Back Azimuth	Difference in Degrees
A to B	_____	_____	_____
B to C	_____	_____	_____
C to D	_____	_____	_____
D to E	_____	_____	_____
E to F	_____	_____	_____

Pacing

Pacing will be one of the methods you use to measure distance. There are more accurate methods. However, it is important to be able to draw rough maps when instruments are not available. Pacing can be moderately accurate except on rough terrain.

The lab instructor has marked a distance of _____ ft or m. Walk from one end to the other counting your paces, or steps. Walk this distance 3 times to determine your average pace.

Number of Paces	Distance per Pace	Average Pace
_____	_____	
_____	_____	
_____	_____	_____ ft or m

Triangulation

Triangulation is a method of locating a point by taking at least two azimuths from known locations, and can be used to determine the distance to a site without direct measurement. To locate an object, measure the azimuth to the object from both ends of a baseline with a known length. The position of the object is determined by the intersection of two azimuths, as shown in Figure 1.2.

When triangulating positions, the angles between the baseline and the line of site of the azimuth should be between 30° and 60°. Smaller angles can be used. However, doing so potentially generates large errors, and therefore, great care must be taken.

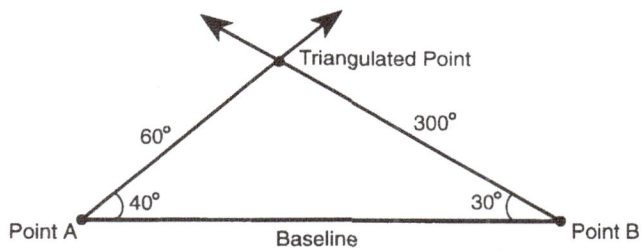

FIGURE 1.2 Determining the position of a point by measuring azimuths from the endpoints of a baseline.

Position   *Azimuth*

First Triangulation

Point _____ to point _____   _____°

Point _____ to point _____   _____°

Second Triangulation

Point _____ to point _____   _____°

Point _____ to point _____   _____°

Mapping Exercise

The lab instructor will assign an area to be mapped and will establish a baseline.

1. Measure the azimuth between points designated by the lab instructor and then determine the distance between them by pacing.
2. Triangulate the corners of the area to be mapped (and other features) designated by the lab instructor.
3. Record your notes, measurement, and sketches on the following page.
4. After completing the field survey, plot the features on the graph paper at the end of the exercise using a scale of 1: _____.

On the graph paper, all vertical lines will be oriented north-south, and the horizontal line will be east-west. Draw the map with a lead pencil, not a pen, in case corrections are needed. Begin with Point 1 and proceed in the order data was collected. The quality of your map depends on accuracy! Use an engineer's scale or ruler for all line work, and a protractor to determine angles.

The map must include a title, scale, date, location, and the azimuth and distance between all points.

FIELD NOTES

From Point	To Point	Azimuth	Paces	Distance	Comments

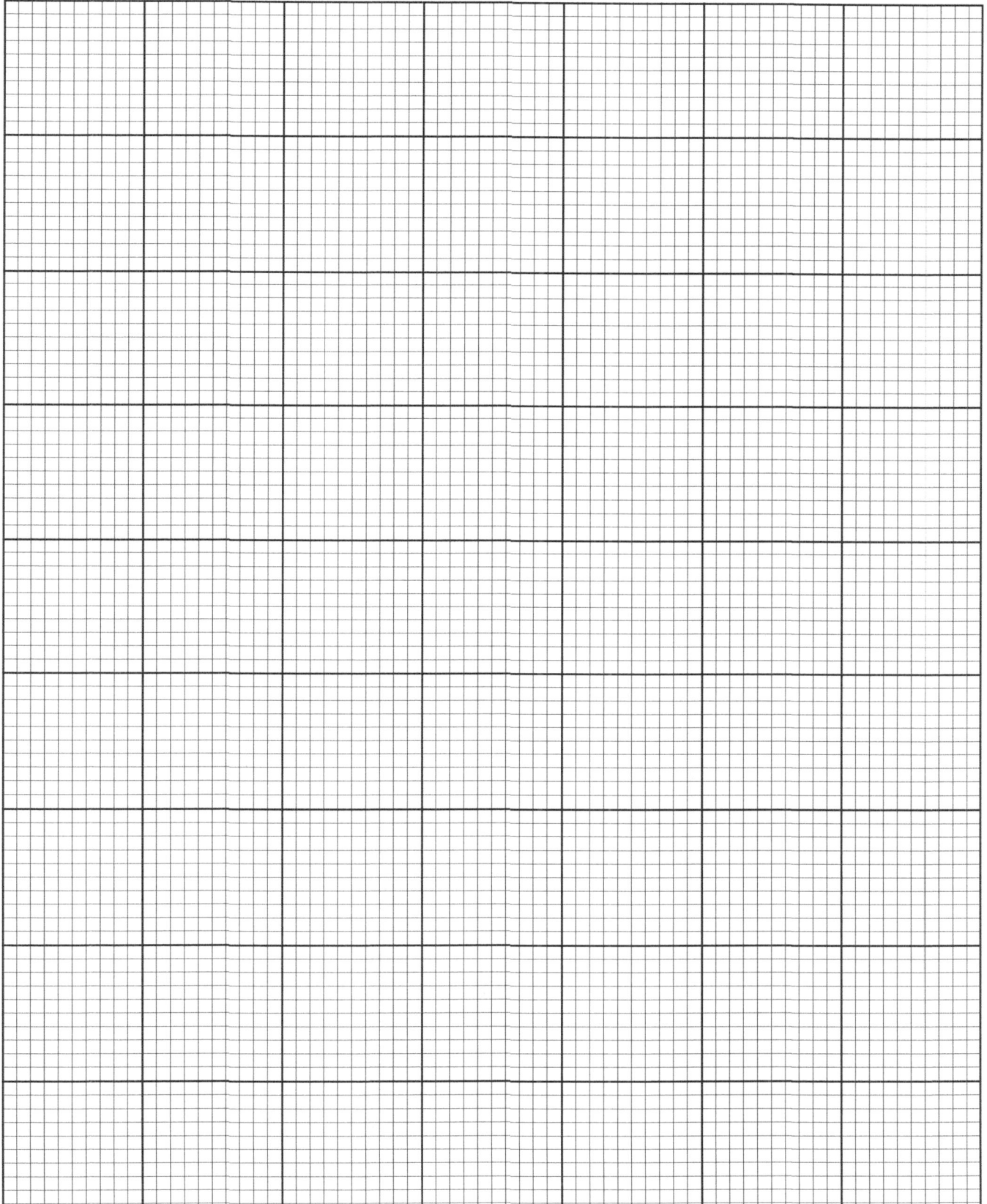

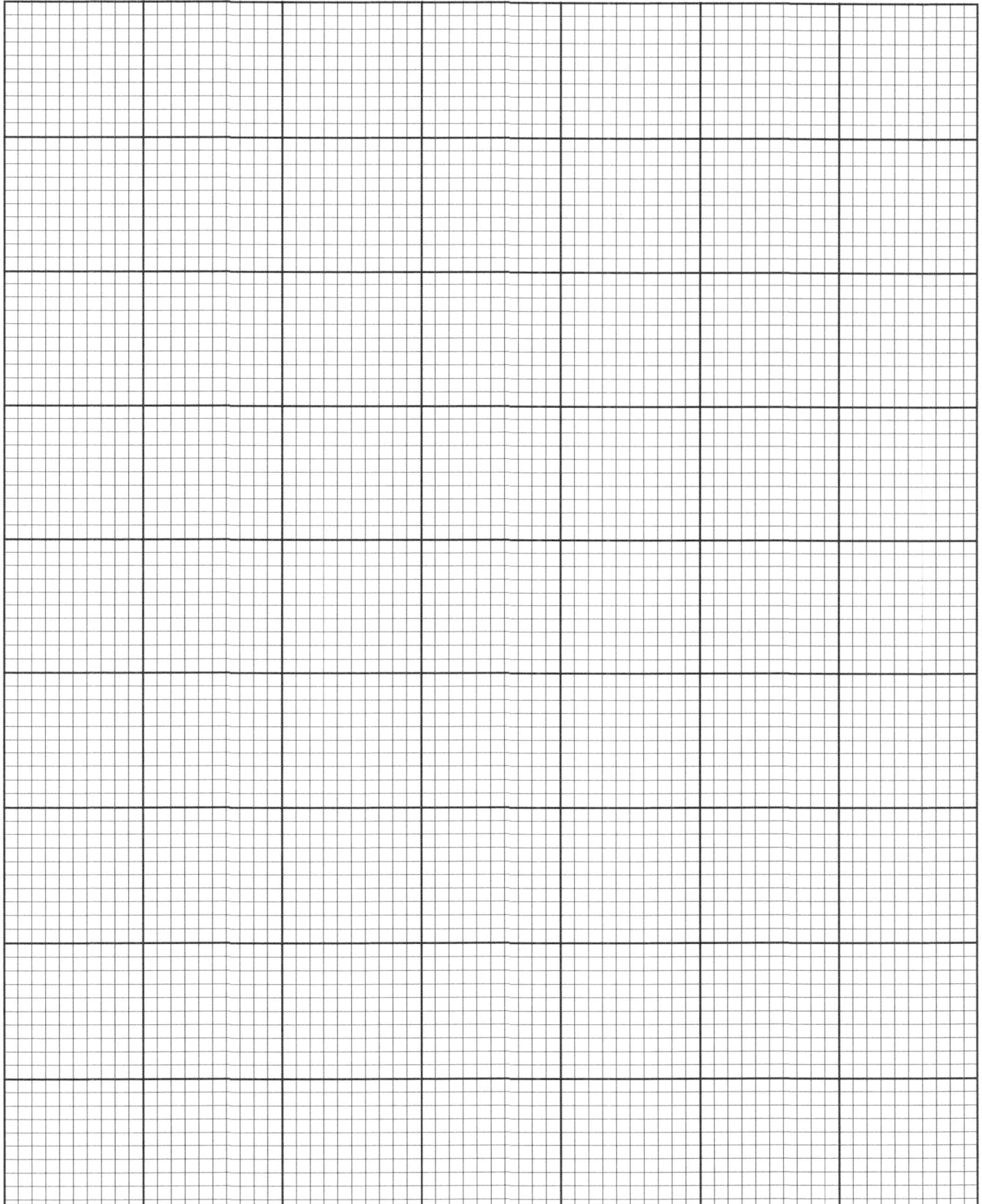

Name _____ Section _____

LAB EXERCISE

Latitude, Longitude, and Time

2

A **geographic grid** is a series of intersecting lines used to locate positions on maps. Most maps are flat and use horizontal and vertical lines. Typically the lines are identified by number and the intersection of lines can be used to indicate a position. The purpose of this exercise is (1) to become familiar with latitude and longitude, the most commonly used geographic grid system to locate positions on the earth's surface, and (2) to understand the relationship between time, longitude, and the earth's rotation.

Latitude and Longitude

Latitude and longitude is a system of imaginary intersecting north-south and east-west lines on the earth's surface.

Latitude is a measure of the distance, or degrees north or south, from the **equator** (Lat. 0°). To determine latitude, there are a series of east-west lines that circle the earth known as **parallels**. The highest latitudes are at the **North Pole** (Lat. 90°N) and **South Pole** (Lat. 90°S) which are fixed points at the earth's axis.

Longitude is a measure of the distance, or degrees, east or west. **Meridians** are north-south lines that are used to measure longitude. The **prime meridian** (Long. 0°) runs through Greenwich, England. All other points of longitude are measures of degrees either east or west of the Prime Meridian, with a maximum of 180° located in the Pacific Ocean. Figure 2.1 illustrates parallels and meridians on the earth's surface.

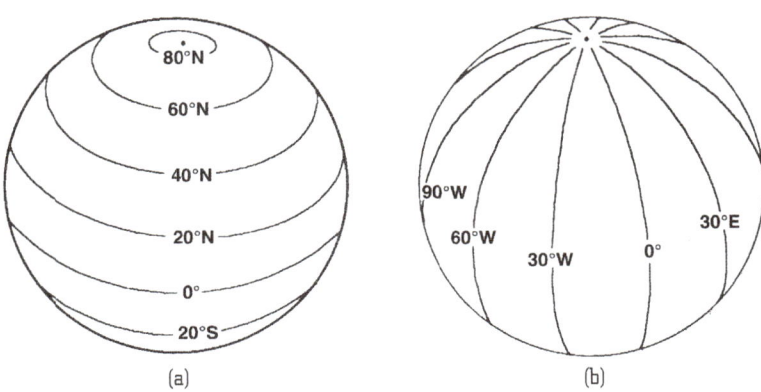

FIGURE 2.1 Parallels (a) and meridians (b).

9

On the Figure 2.2 below:

1. Plot and number the following coordinates:
   a. Lat. 44°N, Long. 65°W
   b. Lat. 12°S, Long. 4°E
   c. Lat. 50°N, Long. 60°E

2. Sketch and label the following parallels:
   a. Arctic Circle
   b. Tropic of Cancer
   c. Tropic of Capricorn

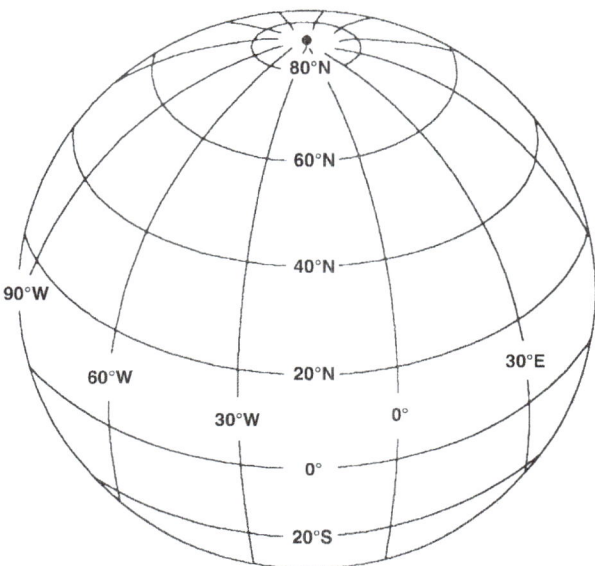

FIGURE 2.2 Geographic grid coordinates.

The earth is often depicted on flat maps because they are easier to work with than globes. The surface of a sphere, however, cannot be flattened without some distortion of the surface features shown. The Mercator Projection (Figure 2.3) is one of the most common maps of the earth. Note on this map that the meridians are parallel. Actually, meridians are farthest apart at the equator and converge at the poles. Therefore, this map projection causes distortion, particularly at the high latitudes. Still, you can easily use grid coordinates to locate positions on this map.

1. Find your position on the Mercator Projection (Figure 2.3) and determine the geographic grid coordinates.

   Lat. _____, Long. _____

2. The following positions will be given by the lab instructor. Locate and plot each position on the Mercator Projection (Figure 2.3).

	Geographic Grid Coordinates	Place Name
a.	_____	_____
b.	_____	_____
c.	_____	_____
d.	_____	_____
e.	_____	_____

3. Shade in the triangular area of the Mercator Projection bounded by:

   Lat. 20°S, Long. 15°E
   Lat. 35°N, Long. 75°E
   Lat. 10°S, Long. 117°E

4. What are the geographic grid coordinates of the positions diametrically opposite the following?

   a. Lat. 18°N, Long. 50°W: _____

   b. Lat. 61°S, Long. 11°W: _____

   c. Lat. 44°, 16'S, Long. 152°, 57'E: _____

   d. _____ : _____

FIGURE 2.3 Mercator Projection. *(Produced by the Cartographic Research Lab, University of Alabama.)*

Latitude and Longitude Distances

The length of one degree of latitude on the earth's surface is approximately 111 km. This varies only slightly with distance from the equator because the earth is not a perfect sphere. In contrast, the circumference of parallels (and one degree of longitude) decreases as you move from the equator (Lat. 0°) to the North Pole or South Pole (Lat. 90°N or 90°S). Longitudinal distance can be easily calculated for any latitude. The circumference of the earth (C) at a given latitude (parallel) decreases in direct proportion to the cosine of the angle of latitude (Figure 2.4). The circumference of a parallel of latitude (distance in kms) can be calculated as follows:

$C = 2 \text{ pi } r \cos (x)$,

where pi = 3.14159..., r = the earth's equatorial radius = 6378 km (3960 miles), and x = the latitude.

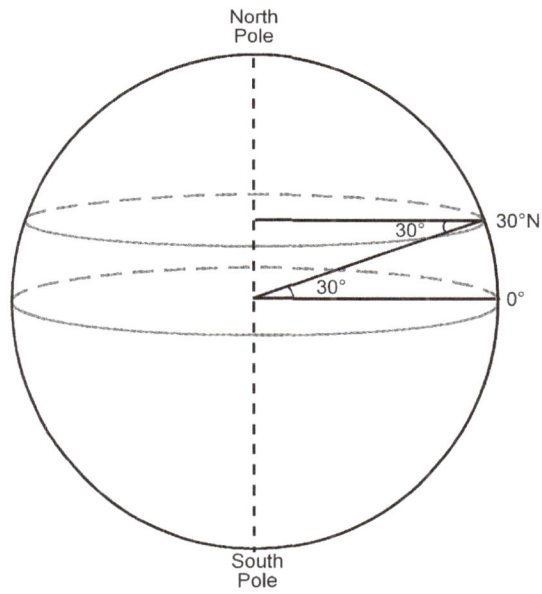

FIGURE 2.4 Angular difference between the Equator and the 30°N parallel.

The length of one degree of longitude at the given latitude parallel = C / 360.

This formula is not entirely correct because the earth is not a perfect sphere. The earth bulges at the equator, and therefore, the equatorial diameter is slightly greater than the polar diameter. However, this formula is accurate to within one percent.

Determine distances for the given latitudes:

	Latitude	Parallel Circumference	1° of Longitude
a.			
b.			
c.			
d.			

Determining Distance by Coordinates

Distances between points within a small region—a few degrees latitude and longitude—can be easily calculated, even if the points are not on the same parallel or meridian. The calculation uses the Pythagorean Theorem that states in a right triangle:

$$c^2 = a^2 + b^2,$$

where c is the distance between the points (hypotenuse), and a and b are the adjacent sides of the triangle that follow a meridian and parallel, as shown in Figure 2.5.

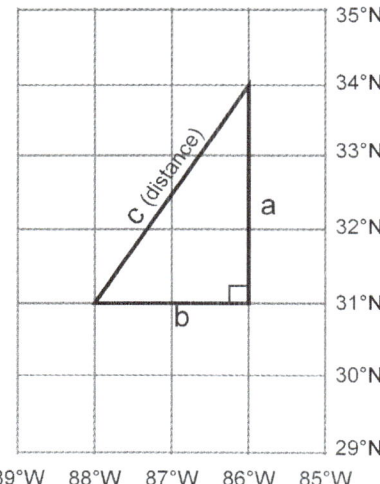

FIGURE 2.5 Map coordinates showing a right triangle.

To conveniently find the distance between 2 points we can restate this simple formula, so that

$$c = \sqrt{(x_2 - x_1)^2} + \sqrt{(y_2 - y_1)^2}$$

where x is the latitudinal distance and y is the longitudinal distance.

Remember, 1° of latitude is constant (111 km). However, you must determine the distance of 1° of longitude for the given latitude as shown on page 14.

Use the graph paper at the end of this exercise to plot points, designated by latitude and longitude, for the following problems.

Problem 1

Point A   Lat. _____   Long. _____

Point B   Lat. _____   Long. _____

Latitudinal distance between points (x): _____

Longitudinal distance between points (y): _____

Distance between points (d): _____

Problem 2

Point C   Lat. _____   Long. _____

Point D   Lat. _____   Long. _____

Latitudinal distance between points (x): _____

Longitudinal distance between points (y): _____

Distance between points (d): _____

Problem 3

Point E   Lat. _____   Long. _____

Point F   Lat. _____   Long. _____

Latitudinal distance between points (x): _____

Longitudinal distance between points (y): _____

Distance between points (d): _____

Great Circle Distances

A great circle is the largest circle possible on a sphere. It is defined by the intersection of a sphere and a plane that passes through the center of the sphere. Great circles bisect the sphere into two equal hemispheres. An arc of a great circle is the shortest distance between two points on the surface of the sphere (Figure 2.6). The equator is a great circle and all meridians are arcs of great circles. An infinite number of great circles are possible by passing a plane through the center of the earth at different angles relative to the equator.

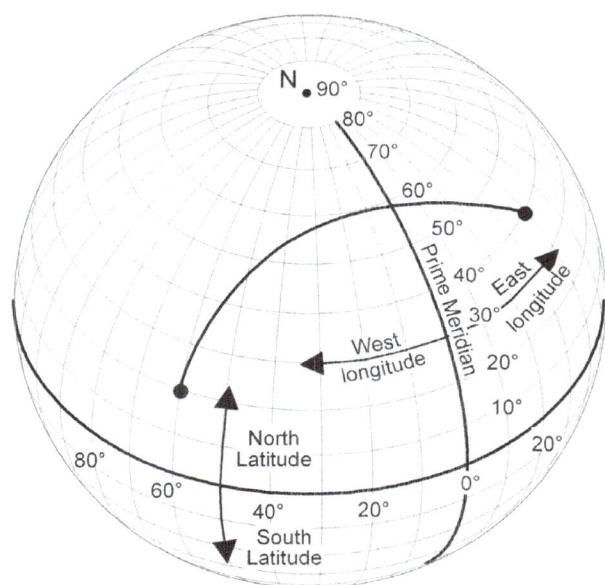

FIGURE 2.6 Map showing an arc of a great circle crossing the Prime Meridian.

Calculating distances using coordinate systems and the Pythagorean Theorem can be used for small areas as previously shown. However, this method cannot be used for large areas—more than a few degrees latitude and longitude—because it does not take into account the curvature of the earth's surface. The distance of an arc of a great circle can be easily calculated given the latitude and longitude of the two points using the following formula:

$$\cos(d) = (\sin a)(\sin b) + (\cos a)(\cos b)(\cos p)$$

where:

      d is the angular distance between points A and B
      a is the latitude of point A
      b is the latitude of point B
      p is the shortest longitudinal difference between points A and B

Note that once you determine cos (d), and angular distance between the two points (d) is determined by using the ARCOS function.

Example:

Calculate the distance between New York City (40°N, 74°W) and Los Angeles located (34°N, 118°W). In this problem, New York City is point A and Los Angeles is point B. Substituting latitude and longitude in the above equation we have:

cos (d) = (sin 40) (sin 34) + (cos 40) (cos 34) (cos 44)
cos (d) = (0.642) (0.559) + (0.766) (0.829) (0.719)
cos (d) = 0.358 + 0.456 = 0.814

d = arcos (0.814) = 35.51

Multiply the angle of degrees (35.51) separating New York City and Los Angeles by 111 km.

Distance = 35.51 * 111 = 3940.6 km (2449 miles)

Calculate the distance between the following points:

Problem 1

a. Atlanta, Georgia (34°N; 84°W)
b. Chicago, Illinois (42°N; 88°W)

    d (angular distance): _____

    distance (km): _____

Problem 2

a. Chicago, Illinois (42°N; 88°W)
b. London, England (52°N; 0°W)

    d (angular distance): _____

    distance (km): _____

Problem 3

a. _____ (Lat. _____ ; Long. _____)

b. _____ (Lat. _____ ; Long. _____)

    d (angular distance): _____

    distance (km): _____

Time

The time of the day is determined by the position of a meridian relative to the sun. It is **noon** on the meridian facing the sun and **midnight** on the diametrically opposite meridian. The earth rotates 360° per day or 15° per hour. The earth rotates west-to-east, and therefore, positions to the east are always later and positions to the west are always earlier. Because the earth rotates, the noon meridian (and the time at all meridians) is constantly changing.

Most of our time keeping uses **standard time zones** that are approximately 15° of longitude wide. All areas within a time zone have the same time of day, which eliminates the need to constantly change times as you move east or west (if you remain in the same time zone). There are 24 time zones, each 15° of longitude and representing one hour of the day. Adjacent time zones differ by one hour. The time zone immediately to the east is one hour later and to the west is one hour earlier.

The time zones are centered on the **central** or **standard meridians** of the zones, which are at 15° intervals beginning with the prime meridian: 0°, 15°, 30°, 45°,...180°. The boundaries of the time zones extend 7½° on either side of the central meridian. The exact boundaries of the time zones, however, are irregular so that they coincide with political boundaries. Figure 2.7 illustrates the world time zones.

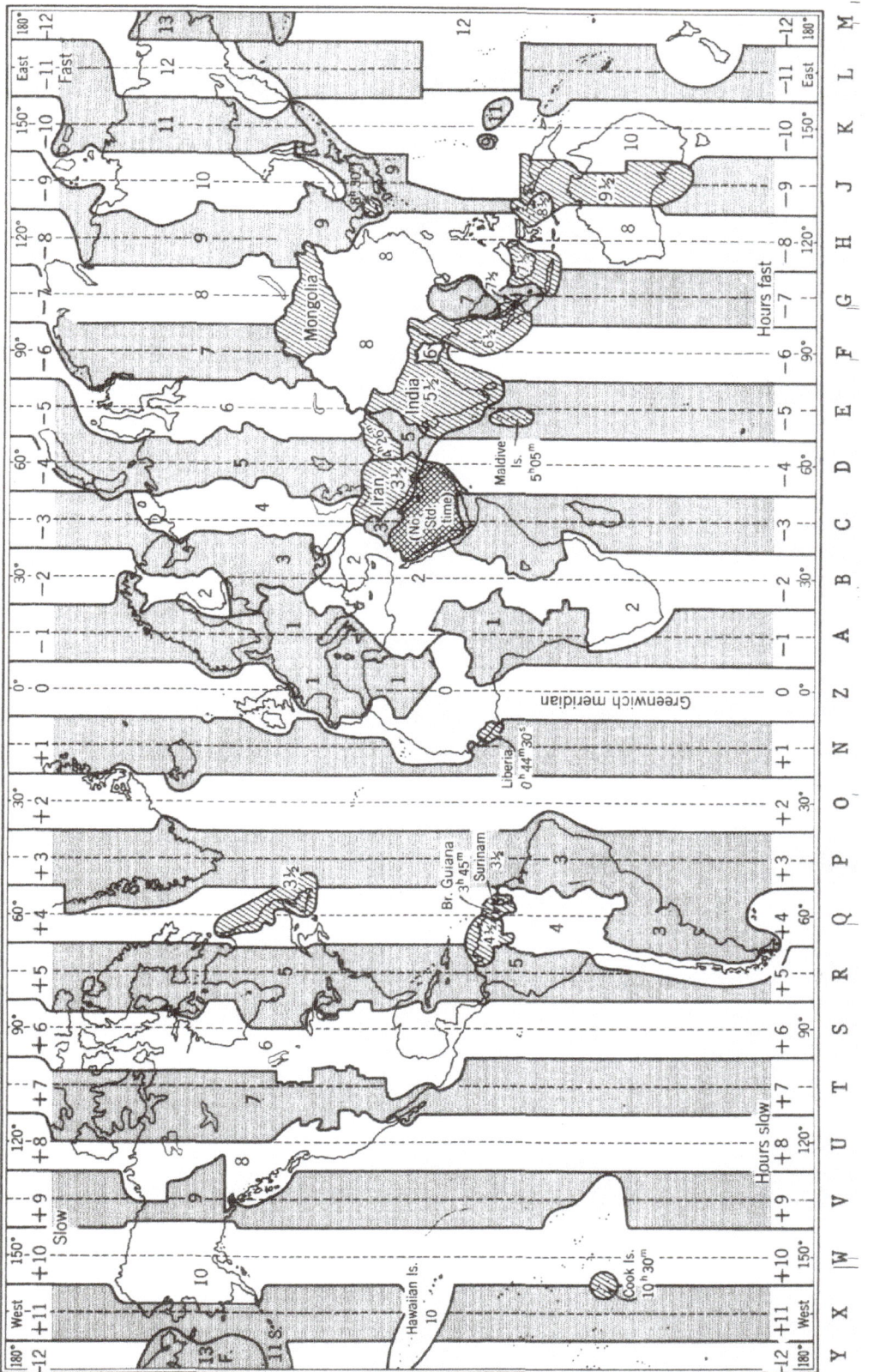

FIGURE 2.7 World time zones. (After U.S. Navy Oceanographic Office, No. 5192.)

19

On the Mercator Projection (Figure 2.8):

1. Mark the central meridians and time zone boundaries for zones in the U.S. Although the boundaries are irregular, assume they are straight in this exercise.

2. Mark the central meridian and time zone boundaries for Greenwich time.

3. Based on your longitude of _____, what is the central meridian of your time zone? _____

4. If the time is _____ in your time zone, what is the time (standard time) at the following meridians?

Longitude	Hour
0°	_____
10°E	_____
_____	_____
_____	_____
_____	_____
_____	_____

5. If it is noon at the following meridians, what is the time in your time zone?

Longitude (noon)	Your Time
0°	_____
33°E	_____
52°W	_____
100°E	_____
_____	_____

FIGURE 2.8 Mercator Projection. *(Produced by the Cartographic Research Lab, University of Alabama.)*

International Date Line

Each new day begins at the international date line. When calculating time for different longitudes, remember to change days when crossing the midnight meridian or the international date line. The international date line coincides with the 180° meridian which is opposite the Prime (Greenwich) Meridian (Figure 2.9).

FIGURE 2.9 International date line.

The rules for changing days when crossing the international date line are:

1. When crossing the international date line moving to the west, you move into the next (new) day (Tuesday into Wednesday). All positions west of the international date line until reaching the midnight meridian are part of the new day.
2. When crossing the international date line moving to the east you move into the previous (old) day (Wednesday becomes Tuesday). All positions east of the international date line until reaching the midnight meridian are part of the same day.
3. You must change days whenever crossing the midnight meridian or international date line, unless it is midnight on the international date line. In this case, the entire earth is experiencing the same day.

Below are pairs of locations. You are given the time and day for one and are to determine the time and day for the corresponding location. Use the Mercator Projection to find the coordinates of each position.

		Hour	Day		Hour	Day
1.	New York, NY	10:24 p.m.	Wed.	Los Angeles, CA		
2.	Denver, CO	2:30 a.m.	Fri.	Tokyo, Japan		
3.	Greenwich, England	8:30 p.m.	Tues.	Tehrān, Iran		
4.	Bombay, India	6:18 a.m.	Wed.	Seattle, WA		
5.	Toronto, Canada			Cairo, Egypt		

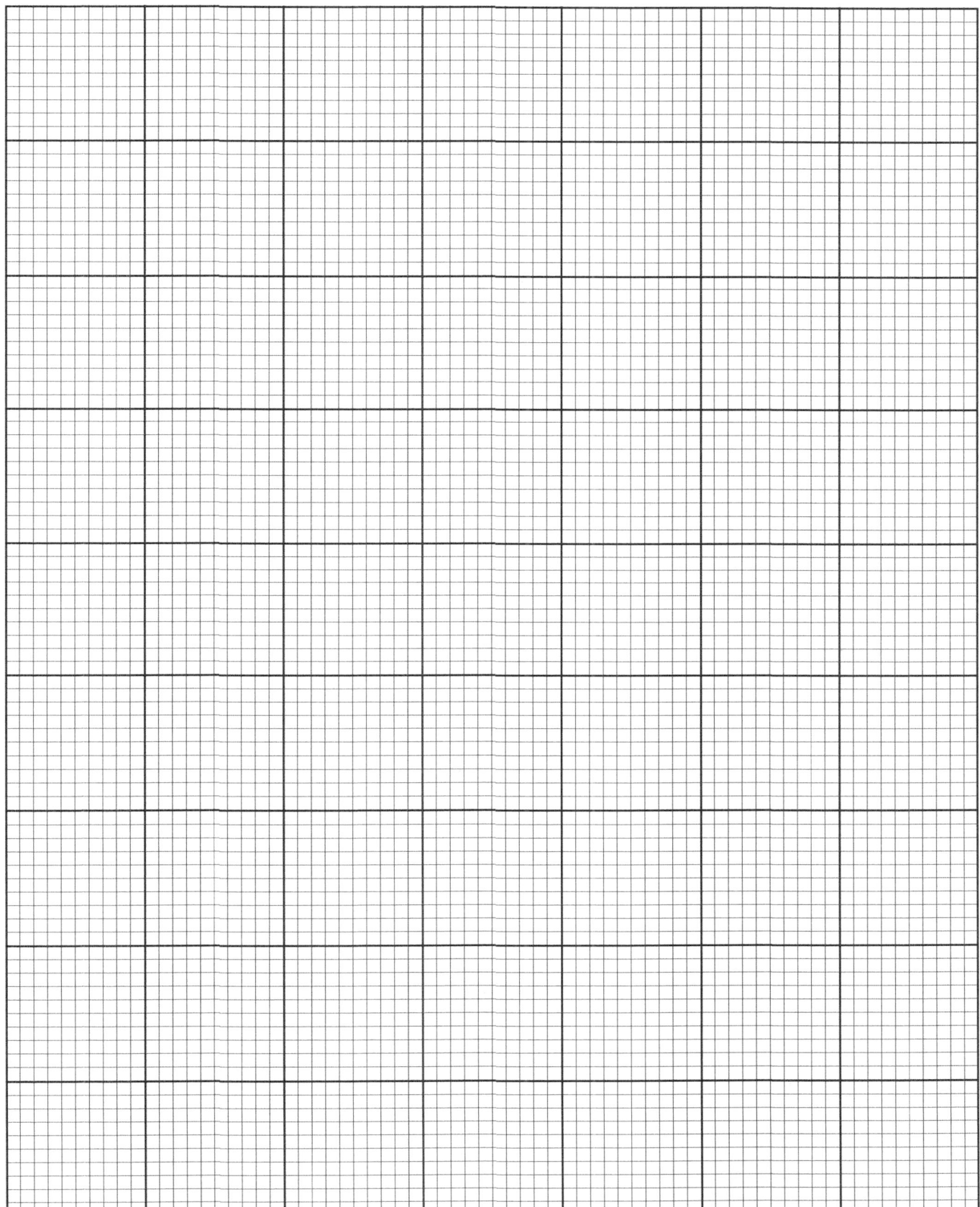

Name _____ Section _____

LAB EXERCISE 3

Earth-Sun Relations

Understanding the earth's movement in space and its changing relationship with the sun is essential for understanding atmospheric and earth-surface processes. The earth's movement directly affects the intensity and duration of solar radiation which accounts for the seasonal changes in temperature and precipitation. Seasonal climatic changes directly affect vegetation, hydrology, soil formation and erosion, ocean currents, among other processes. The purpose of this exercise is to develop an understanding of (1) earth rotation and revolution and (2) changing declination of the sun and how this affects changes in the duration and intensity of solar radiation.

Earth Rotation

Rotation refers to the motion of the earth turning on its axis. The direction of rotation is west-to-east, or counter-clockwise as you look down on the North Pole. A **solar day** is defined as one complete rotation relative to the sun.

While the earth's **angular velocity** is 360°/day (15°/hour), its speed, or **linear velocity**, depends on latitude. The distance around the earth at the equator is 24,900 miles (40,075 km). A point at the equator would move this distance each day. What is the earth's speed at the equator (Lat. 0°)?

_____ miles/hour

_____ km/hour

As shown in Figure 3.1, the distance around each parallel of latitude decreases with distance toward the Poles. Therefore, while every position on earth rotates 360°/day, the distance traveled, and therefore speed, depends on latitude.

27

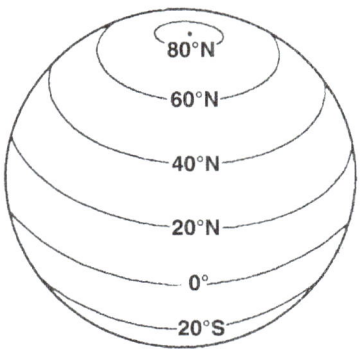

FIGURE 3.1 Parallels of latitude.

Rotational velocity can be calculated for any latitude by dividing the length of a parallel (distance around the earth at that latitude) by 24, the number of hours in a day. Calculate the rotational velocity for each latitude shown below.

Latitude (N or S)	Length of 1° Longitude		Total Length		Rotational Velocity	
	miles	km	miles	km	mph	kmph
20	65	105	_____	_____	_____	_____
40	53	85	_____	_____	_____	_____
60	35	56	_____	_____	_____	_____
80	12	19	_____	_____	_____	_____

What is the rotational velocity of:

North Pole (Lat. 90° N)? _____

South Pole (Lat. 90° S)? _____

Declination of the Sun

The earth revolves in its orbital path around the sun on the plane of the ecliptic. The earth's axis is inclined 23½° to a line perpendicular to the ecliptic (or 66½° from the ecliptic). The earth's axis is in a fixed position relative to the stars, so that the North Pole (Lat. 90°N) always faces the North Star.

As the earth moves in its orbital path, the axis changes its position relative to the sun. (Figure 3.2.) Because of the earth's 23½° axis tilt, the direct ray of the sun, or the latitude on earth where the sun is directly overhead, moves between the Tropic of Cancer (Lat. 23½°N) and the Tropic of Capricorn (Lat. 23½°S). The direct ray of the sun, or declination, for every day of the year is shown on the analemma (Figure 3.6) at the end of this exercise. Note that the direct

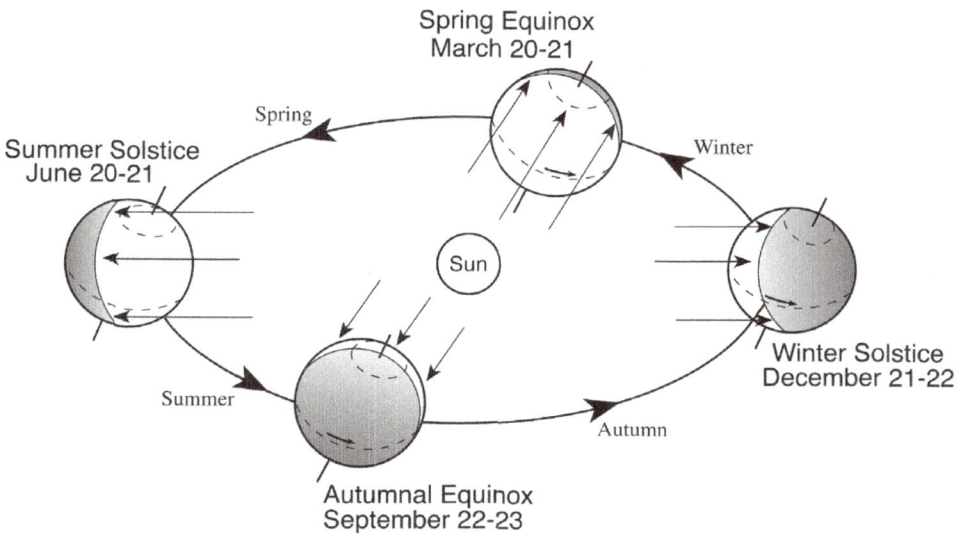

FIGURE 3.2   Earth's revolution and axis position relative to the sun.

ray of the sun is farthest north in late June, which is the **summer solstice**, and farthest south in late December, the **winter solstice**. The sun crosses the equator (Lat. 0°) twice each year on the **March** and **September equinoxes**.

Refer to the analemma (Figure 3.6) for the following questions:

1. What is the declination of the sun?

    today: _____°

    one week from today: _____°

    one month from today: _____°

    six months from today: _____°

2. Between what two dates is the declination of the sun

    moving north? _____

    moving south? _____

3. Between what two dates is the declination of the sun

    in the northern hemisphere? _____

    in the southern hemisphere? _____

Length of Day and Night

Because the earth is a sphere, one half (the part facing the sun) is illuminated at any point in time. However, because the position of the axis is changing relative to the sun, the amount of daylight, as opposed to darkness, changes at all positions or latitudes, except at the equator which always has 12 hours of daylight and darkness. Figure 3.3 shows the earth-sun relationships on the June (summer) and December (winter) solstices.

Rules for the length of day and night

1. If the declination of the sun is in the northern hemisphere, the length of day is greater than night for all positions in the northern hemisphere, and length of night is greater than day for all positions in the southern hemisphere.

   Conversely, if the declination of the sun is in the southern hemisphere, the length of day is greater than night for all positions in the southern hemisphere, and length of night is greater than day for all positions in the northern hemisphere.

2. If the declination of the sun is in the northern hemisphere, there is progressively longer period of daylight with distance north, until reaching the area receiving 24 hours of daylight.

   If the declination of the sun is in the southern hemisphere, there is progressively longer period of daylight with distance south, until reaching the area receiving 24 hours of daylight.

3. All parallels of latitudes will have either greater periods of daylight or darkness for each 24 hour period except on an equinox when all parallels have 12 hours day and night. The exception is the equator (Lat. 0°) which always has 12 hours of daylight and darkness.

4. As the declination of the sun moves north (from the December solstice to the June solstice) the period of daylight increases in the northern hemisphere and decreases in the southern hemisphere.

   As the declination of the sun moves south (from the June solstice to the December solstice) the period of daylight increases in the southern hemisphere and decreases in the northern hemisphere.

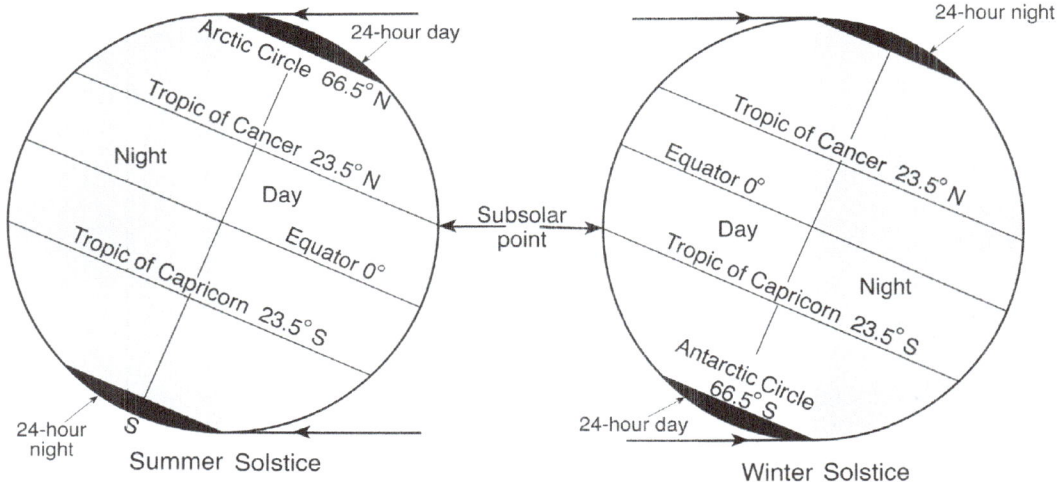

FIGURE 3.3 Illumination of the earth on the solstices.

Refer to Figure 3.3 to answer questions 1 and 2.

1. Compare the length of daylight at the tropic of Cancer (Lat. 23.5°N) on the June and December solstices.

2. Compare the length of daylight at the tropic of Capricorn (Lat. 23.5°S) on the June and December solstices.

Questions 3–7 refer to your latitude: _____ ° N or S

3. Today (date: _____) is the period of daylight or darkness longer?

   _____

4. Tomorrow, will the length of day be longer or shorter than today?

   _____

5. On what day of the year will the period of daylight be longest?

   _____

6. On what day of the year will the period of darkness be longest?

   _____

7. During what time of year do you have longer periods of daylight than darkness?

   _____

8. Compare the lengths of daylight today between:

   10°N and 20°N: _____

   20°S and 0°: _____

   _____ and _____ : _____

9. Which of the following latitudes:

   Tropic of Cancer (Lat. 23.5°N)
   Equator (Lat. 0°)
   Tropic of Capricorn (Lat. 23.5°S)

   will have a greater period of daylight on:

   January 1? _____

   May 14? _____

   August 25? _____

   October 1? _____

   _____ ? _____

   _____ ? _____

10. On the following date: _____, is the length of daylight increasing or decreasing at:

    Lat. _____ : _____

    Lat. _____ : _____

    Lat. _____ : _____

    Lat. _____ : _____

There are always areas on earth that have either 24 hours of daylight or darkness (except during an Equinox when all parallels have 12 hour days and nights). These are the regions in the high latitudes that do not rotate either into or out of the **circle of illumination** (or surface area exposed to the sun) as the earth turns on its axis. The areas on earth receiving 24 hours of daylight or darkness can easily be determined for any day of the year by referring to the analemma.

First, find the declination of the sun for the desired date and then subtract it from 90 (90 – declination). The resulting value represents the latitudes at and above which there is either 24 hours daylight or darkness. For example, if the declination of the sun is 10°N (90 – 10 = 80), latitudes 80°N and S and all higher latitudes will have either 24 hours daylight or darkness. Since in this case the sun is in the northern hemisphere, all areas ≥ 80°N will have 24 hours daylight on that day. All areas ≥ 80°S will have no daylight.

Since the maximum declination of the sun is 23.5°N or S, the lowest latitude to have 24 hours of daylight and darkness at some point during the year is Lat. 66.5°N (**Arctic Circle**) and Lat. 66.5°S (**Antarctic Circle**) (90 − 23.5 = 66.5).

1. What areas will receive 24 hours of daylight?

    today: _____

    one month from today: _____

    June Solstice: _____

    Date: _____ : _____

    Date: _____ : _____

2. What areas will receive 24 hours of darkness?

    today: _____

    one month from today: _____

    December Solstice: _____

    Date: _____ : _____

    Date: _____ : _____

3. Between what two dates will the following positions have 24 hours of daylight?

    North Pole: _____

    80°N: _____

    70°N: _____

Why is the period of continuous daylight decreasing as you move to a lower latitude?

Altitude of Noon Sun

The altitude of the sun is the angle of the sun above the horizon. The sun reaches its highest point in the sky each day at noon. Since the sun is highest in the sky at noon, the intensity of solar radiation is greatest at this time each day. Figure 3.4 shows the altitude of the sun at the solstices and equinoxes.

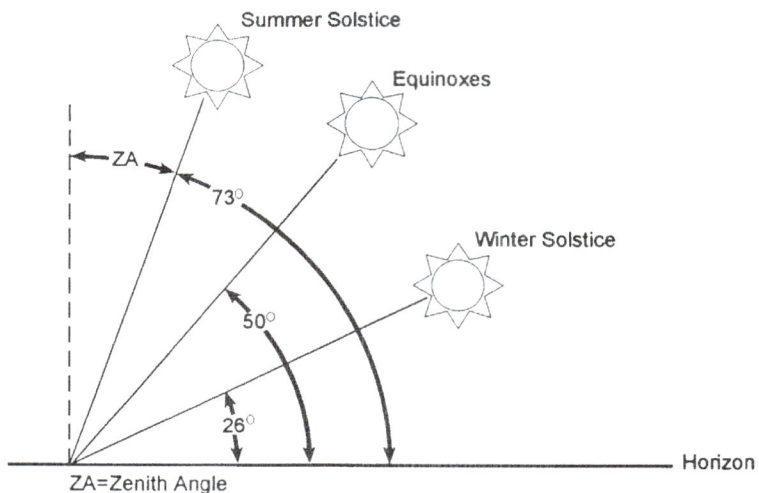

FIGURE 3.4 Altitude of the noon sun at the solstices and equinoxes at 40°N.

The altitude of the noon sun (ANS) will vary depending on latitude. Because the earth is a sphere, the sun can be directly overhead at only one latitude (shown by the analemma) on any day. As the earth moves in its orbital path and the declination of the sun changes, the ANS also changes. If you know your latitude, the ANS can be determined for any day of the year.

The ANS at any latitude can be calculated using the following equation:

ANS = 90 − Zenith, or
ANS = 90 − (arc of earth's surface between LP and LS)

where:
LP is your latitudinal position and
LS is the declination of the sun. Therefore,

$$\text{ANS} = \begin{array}{ll} 90 - (LP+LS) & \text{if LP and LS are in different hemispheres} \\ 90 - |LP-LS| & \text{if LP and LS are in the same hemisphere} \end{array}$$

Note: If the sun is directly overhead, the LP and LS is the same, and therefore, the zenith angle (or difference between LP and LS) is 0. In other words, if the sun is directly overhead, it is 90° above the horizon:

ANS = 90 − 0
ANS = 90

But, for every degree difference between the LP and the LS, the sun is that many degrees from the direct overhead position. For example, as shown in Figure 3.5, the declination of the sun (LS) on May 21 is 20°N. If you were at 20°N (LP), the noon sun would be directly overhead. Now assume that you are at 13°N (LP), or 7° from the declination of the sun (LS). So, instead of the sun being directly overhead, it is 7° from the straight overhead position. The zenith angle is 7°.

ANS = 90 − (20 − 13)
ANS = 90 − 7 = 83°

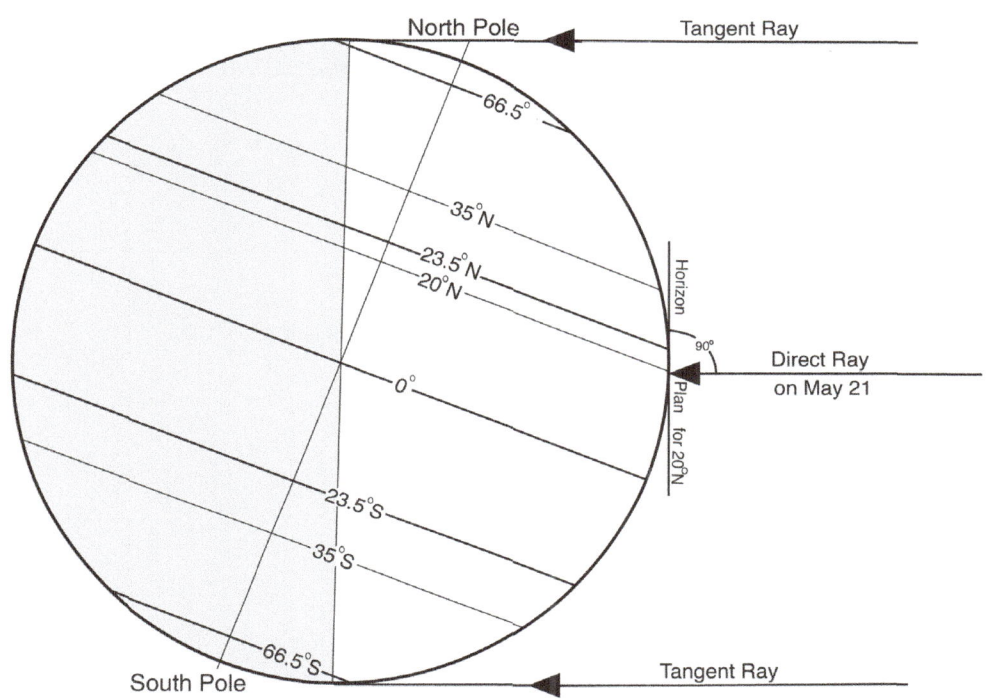

FIGURE 3.5 Earth-sun relationship on May 21 and July 25.

Use your latitude position (LP) for questions 1–5.

Latitude: _____ Today's date: _____

1. The declination of the sun today is _____

2. What is the ANS today at your latitude? _____

3. On what day of the year is the ANS lowest? _____

   What is the ANS on this day? _____

4. On what day of the year is the ANS highest? _____

   What is the ANS on this day? _____

5. What is the ANS on an equinox? _____

6. Today, on what part of the earth's surface does the period of daylight exceed the period of darkness?

7. At what latitudes (in both northern and southern hemispheres) will the sun never be directly overhead?

8. What is the highest and lowest ANS for the following latitudes?

Latitude	ANS Highest	Lowest
_____	_____	_____
_____	_____	_____
_____	_____	_____
_____	_____	_____

Fill in all blanks in the following table:

ANS	LP	LS	Date
_____	10°N	_____	March 28
_____	38°N	_____	October 1
_____	22°S	14°N	_____
_____	0°	2°S	_____
_____	80°N	8°S	_____

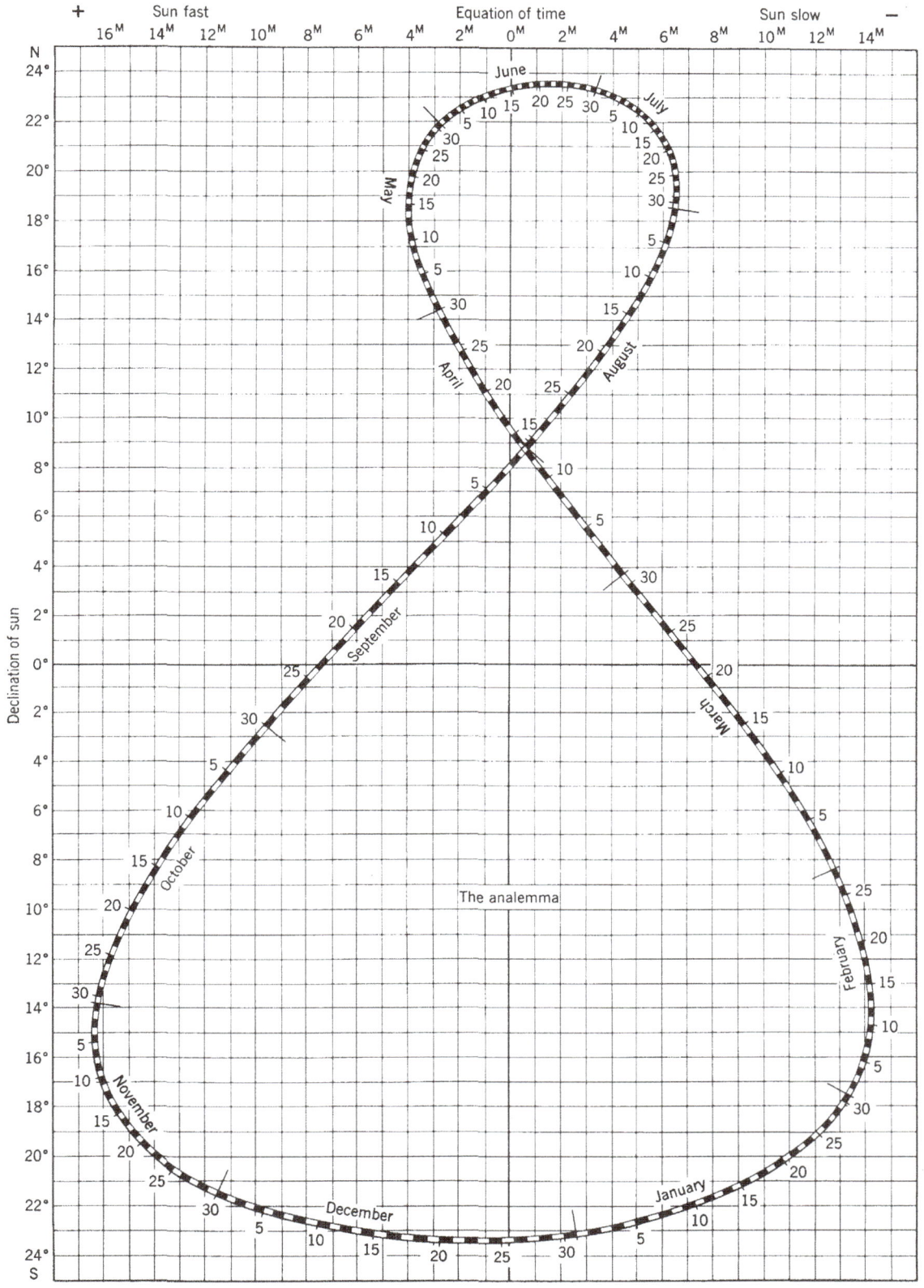

FIGURE 3.6 The analemma gives the declination of the sun and the equation of time for each day in the year. *(From U.S. Coast and Geodetic Survey.)*

Name _____ Section _____

LAB EXERCISE 4

Solar Radiation

Equipment

- pyranometer
- thermometer
- soil thermometer

Solar radiation is the major heat source for the earth's surface and is the power controlling the earth's weather. Solar radiation also, directly or indirectly, affects almost all earth surface processes, including vegetation, soil moisture, stream flow, weathering processes, and ecosystem functioning. In this exercise, data will be collected from different sites to determine how small scale variations in radiation affect temperature. Also, you will calculate the intensity of solar radiation for different latitudes and topographic positions based on the altitude of the sun.

Because of the differences in the altitude of the sun with different latitudes, the intensity of solar radiation at the surface is highly variable. Also, some of the solar radiation reaching the earth's surface is reflected back to space and does not contribute to surface heating. The reflectivity of a surface, or **albedo** (A), is usually expressed as a percentage and can be defined as:

$$A = \frac{\text{reflected solar radiation}}{\text{incoming solar radiation}} * 100$$

The albedo of a surface is highly variable. With identical solar angles, dark or rough surfaces, such as those covered by asphalt or vegetation, typically have low albedos. Whereas, light or smooth surfaces, such as those covered by snow or ice, usually have high albedos. Since reflected solar radiation does not heat surfaces, surface albedo can directly affect both surface and air temperature.

Data Collection

Three sites will be designated from which you will collect radiation and temperature data. A pyranometer will be used to measure both incoming and reflected solar radiation for each study site. The ratio of these two readings will then be used to calculate albedo as shown in the above equation.

On each site, measure the temperatures at the surface, 1 meter above the surface, and at least 8 cm (about 3 inches) below the surface, except on the black asphalt. The subsurface temperature readings should be taken at the same depth on the remaining sites. Record the temperatures and radiation values on the following table. Then calculate albedo.

	Black Asphalt	Vegetated Surface Full Sun	Vegetated Surface Shade	
Temperature (°F)				
1 m above surface	_____	_____	_____	_____
Surface	_____	_____	_____	_____
Soil		_____	_____	_____
Solar Radiation				
Incoming	_____	_____	_____	_____
Reflected	_____	_____	_____	_____
Albedo	_____%	_____%	_____%	_____%

Note: It is best to take temperature readings when there is little wind. Even moderate wind causes turbulence and mixing of the air near the surface. Mixing will reduce or eliminate both vertical and horizontal temperature differences that normally occur when conditions are calm and will limit the ability to make useful interpretations from this data.

On Figure 4.1 plot the three temperature readings for each study site. Then, connect the points to create temperature profiles.

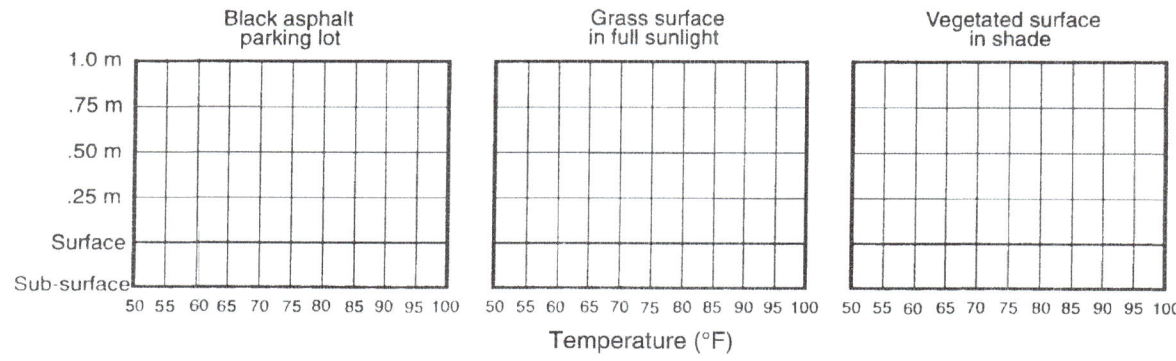

FIGURE 4.1 Create temperature profiles.

1. On which site is albedo greatest? Why?

2. On which site is the temperature difference between the surface and 1 meter elevation greatest? Why?

3. Which surface has the highest temperature and what is the relationship between surface temperature and albedo?

4. Are the soil temperatures similar on the shaded and unshaded sites? Explain.

5. What differences would occur in temperature at each elevation under cloudy conditions?

Intensity of Solar Radiation

The intensity of solar radiation depends on the angle of the solar beam which is determined by latitude and the changing declination of the sun. If the sun is directly overhead, the intensity of radiation is greatest because of its concentration in a small area. As shown in Figure 4.2, with a lower angle, the same amount of radiation is distributed over a larger area and therefore is less intense.

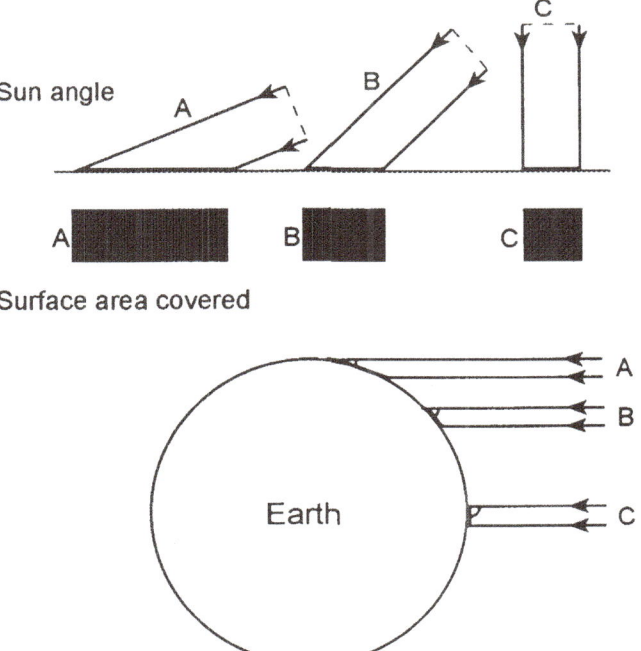

FIGURE 4.2 Angle of the sun's rays and concentration of radiation.

The intensity of radiation for different sun angles is shown in Table 4.1. Radiation intensity for given angles (altitude) of the sun is expressed as a percentage of that of a beam directly overhead (perpendicular to the surface).

Table 4.1 INTENSITY OF SOLAR RADIATION

Angle of Beam	0°	1°	2°	3°	4°	5°	6°	7°	8°	9°
0°	00.0	01.8	03.5	05.2	07.0	08.7	10.5	12.2	13.9	15.6
10°	17.4	19.1	20.8	22.5	24.2	25.9	27.6	29.2	30.9	32.6
20°	34.2	35.8	37.5	39.1	40.7	42.3	43.8	45.4	47.0	48.5
30°	50.0	51.5	53.0	54.5	55.9	57.4	58.8	60.2	61.6	62.9
40°	64.3	65.6	66.9	68.2	69.5	70.7	71.9	73.1	74.3	75.5
50°	76.6	77.7	78.8	79.9	80.9	81.9	82.9	83.9	84.8	85.7
60°	86.6	87.5	88.3	89.1	89.9	90.6	91.4	92.1	92.7	93.4
70°	94.0	94.6	95.1	95.6	96.1	96.6	97.0	97.4	97.8	98.1
80°	98.5	98.8	99.0	99.3	99.5	99.6	99.8	99.9	99.9	99.9

To determine the relative intensity of a solar beam with an angle of incidence of 53 degrees, read down the left-hand column to 50 and then across the row to 3. The relative intensity for an angle of incidence of 53 degrees is 79.9 percent.

For the following problems you must determine the altitude of the noon sun (ANS) for different dates at given latitudes. Refer to the analemma (Figure 3.6) to make this calculation. The latitude of your position is

_____ °N

Use Table 4.1 to answer the following questions.

1. Today, the altitude of the noon sun (ANS) at your latitude is _____ .

   The intensity of solar radiation at noon at your latitude is _____ percent.

2. On what date will maximum intensity of radiation occur at your latitude?

   _____

   What is the altitude of the noon sun (ANS) on that day? _____

   What is the maximum intensity of radiation that occurs on that day?

   _____ percent

3. On what date will minimum intensity of radiation occur at your latitude?

   _____

What is the altitude of the noon sun (ANS) on that day? _____

What is the minimum intensity of radiation that occurs on that day?

_____ percent

On the following table include the ANS and intensity of radiation for the given latitudes.

Latitude	ANS		Intensity of Radiation (%)	
	Highest	Lowest	Maximum	Minimum
0°	_____	_____	_____	_____
10°N	_____	_____	_____	_____
_____	_____	_____	_____	_____
_____	_____	_____	_____	_____
_____	_____	_____	_____	_____
_____	_____	_____	_____	_____

The intensity of radiation over a small area is also affected by topography. In the U.S. the noon sun is always in the southern sky (even though its exact position changes). Therefore, the intensity of radiation is higher on south-facing slopes than on north-facing slopes at the same latitude.

To determine the intensity of radiation on an inclined surface you must determine (1) the solar angle on flat ground at the same latitude, and (2) the angle of the slope. The solar angle on the ground (SA_g) is calculated by:

$$SA_g = SA \pm \alpha$$

where: SA = solar angle on a flat surface
α = slope angle in degrees

Figure 4.3 illustrates how much the intensity of radiation will vary on different facing slopes, even those that are gently inclined.

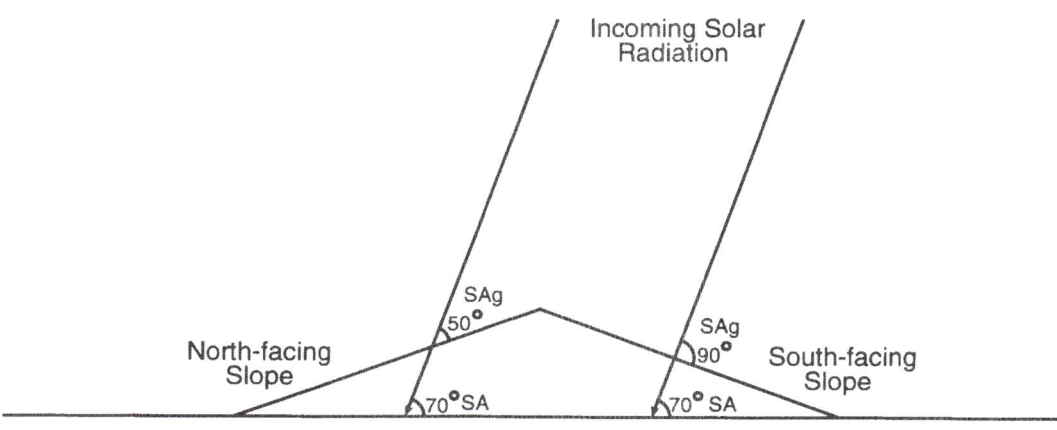

FIGURE 4.3 Angle of the sun on different slopes.

In the following table, slope angles for north and south facing slopes are given. Complete the table by calculating the intensity of solar radiation at noon on slopes with different angles in your area. Refer to Table 4.1.

Latitude: _____ Today's date: _____ ANS: _____

Slope Angle (°)				Intensity of Radiation (percent)	
north-facing	SAg	south-facing	SAg	north-facing	south-facing
5		5			
10		10			

For the following table, dates and latitudinal position (and slope angle where not shown) will be given by the lab instructor. Based on this information, complete the table by calculating the ANS and the intensity of radiation.

Date	Latitude	ANS	Slope Angle (°)		Intensity of Radiation	
			north-facing	south-facing	north-facing	south-facing
			5	5		
			10	10		

Name _____ Section _____

Changing Earth-Sun Relations: Milankovitch Cycles

LAB EXERCISE 5

In Exercise 4, we learned that Earth-Sun geometry is an extremely important factor that influences the amount and intensity of radiation that the Earth receives. Because of this relationship, different locations on Earth receive unequal amounts of radiation, and therefore experience different average temperatures. Up to this point, however, we have assumed that this Earth-Sun geometry is constant and unchanging. We know that this is not the case, as Earth experiences dramatic natural shifts in temperature, resulting in ice ages and interglacial periods.

In the early 20th century a Serbian astrophysicist and mathematician named **Milutin Milankovitch** developed a mathematical theory that attempted to explain natural climate change on Earth. We have known for some time that the Earth's average temperatures are in a constant state of fluctuation. These fluctuations occur on a variety of time scales, and range from very long-term (e.g., the fluctuation between ice ages and interglacials) to very short-term (e.g., the seasonal climate changes associated with El Niño). However, Milankovitch was only concerned with the long-term climate fluctuations that Earth experiences, and proposed that natural changes (cycles) occur over time that affect the Earth-Sun geometry. These changes in geometry affect the amount of solar radiation that the Earth receives, as well as the angle that the energy is received. Over long time scales, these changes in the Earth-Sun relationship can lead to long-term changes in the average temperature on Earth.

Milankovitch proposed three cycles that affect the Earth-Sun geometry:

1. The Earth's orbital pattern around the Sun changes through time. This cycle is known as the **eccentricity cycle**.

2. The **obliquity cycle** states that the tilt of the Earth changes over time.

3. The direction or orientation of the Earth's axis changes over time. This cycle is also known as the **precessional cycle**.

The Eccentricity Cycle

Over the past 2 million years, Milankovitch states that the shape of the Earth's orbit around the Sun has varied from a 0% ellipse (a perfect circle with the Sun directly in the center) to a 6.7% ellipse, as shown in Figures 5.1 and 5.2.

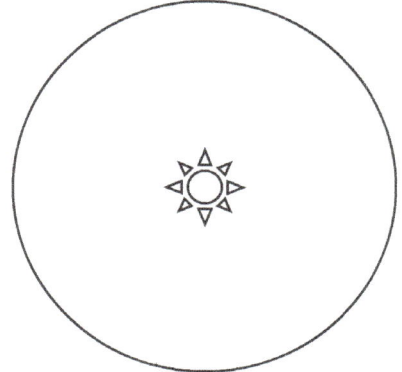

FIGURE 5.1 The Earth's orbit at a 0% ellipse.

FIGURE 5.2 The Earth's orbit at a 6.7% ellipse.

Milankovitch also states that the eccentricity cycle has a periodicity of 100,000 years (Figure 5.3). This does not mean that the Earth's orbit changes from 0% to 6.7%, then back to 0% in 100,000 years. It only means that the wavelength of the cycle (length of time between peaks) is approximately every 100,000 years. Note in Figure 5.9, that in the past 1,000,000 years the eccentricity value has not come close to its 6.7% maximum.

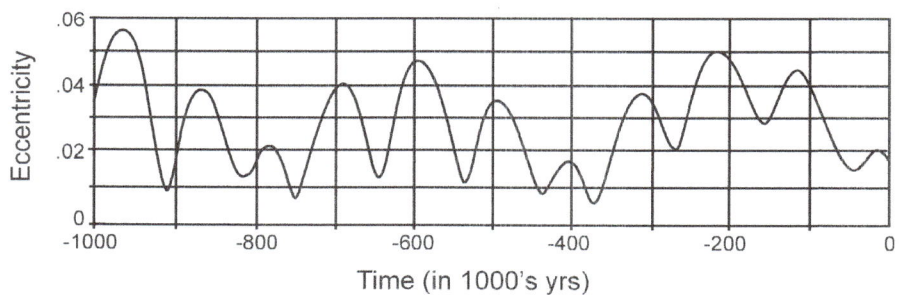

FIGURE 5.3 The eccentricity cycle over the last 1,000,000 years. *(Source: Martin Williams, David Dunkerley, Patrick De Deckker, Peter Kershaw and John Chappell. 1998. Quaternary Environments. Second Edition. Arnold Press, London.)*

How many times in the past 1,000,000 years has Earth's orbit been at the following eccentricity values.

1. 0% _____

2. 2% _____

3. 3% _____

4. 5% _____

5. What other patterns do you notice in the eccentricity cycle, besides the obvious 100,000 year cycle? Hint: focus on every fourth eccentricity cycle.

Since the Earth revolves around the Sun, the Sun is known as the **focus** of the ellipse. When the Earth's orbit is a perfect circle (0% ellipse) then the **focus** is also the **center** of the ellipse. However, as the orbit becomes more elliptical, the Sun migrates from the center of the ellipse (Figure 5.4).

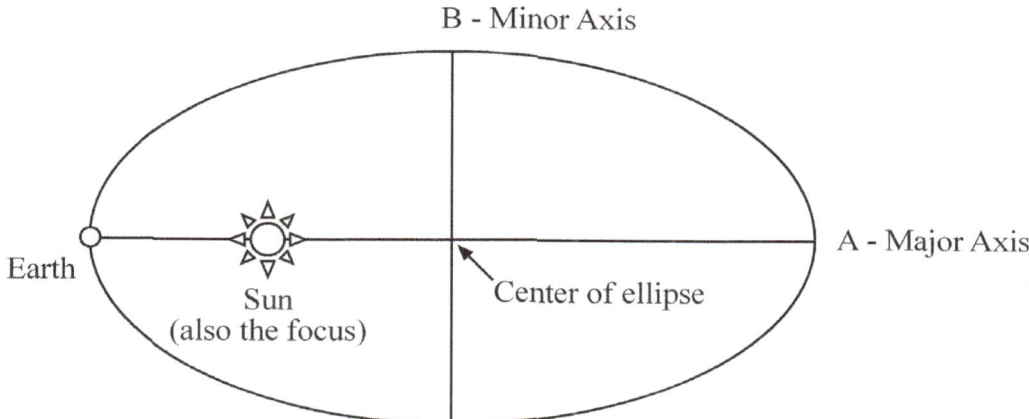

FIGURE 5.4 The Earth revolving around the Sun on an elliptical orbit.

The eccentricity of the Earth's orbit can be calculated with the following equation:

$$E = (\text{distance between F and C}) / A$$

where:

E = eccentricity; F = focus; C = center; A = length of major axis

Calculate the following eccentricities, using these hypothetical values:

6. F = 31; C = 20; A = 100 _____

7. F = 22; C = 20; A = 100 _____

8. F = 18; C = 12; A = 125 _____

9. F = 5; C = 5; A = 40 _____

10. F = 6; C = 5; A = 45 _____

11. F = 70; C = 90; A = 250 _____

12. Which of the questions above (questions 6–11) have eccentricity values that could have been possible in the past 2 million years? _____

13. How could a change in eccentricity affect the amount and seasonality of energy that the Earth receives from the Sun?

The Obilquity Cycle

Over the past 2 million years, Milankovitch states that the tilt of the Earth has varied from roughly 22° to 24.5° (Figure 5.5). Currently the Earth is tilted at approximately 23.44°, which causes the Tropics of Cancer and Capricorn to fall at these latitudes north or south of the equator, respectively.

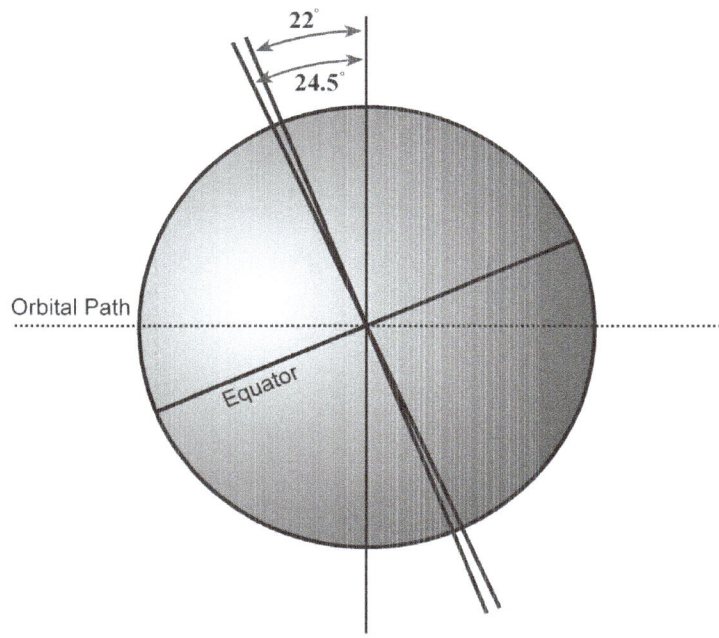

FIGURE 5.5 Variability of the Earth's axis according to the obliquity cycle.

Milankovitch also states that the obliquity cycle has a periodicity of approximately 41,000 years (Figure 5.6). Again, this does not mean that the Earth's tilt changes from 22° to 24.5°, then back to 22° in 41,000 years. It only means that the wavelength of the cycle is approximately every 41,000 years.

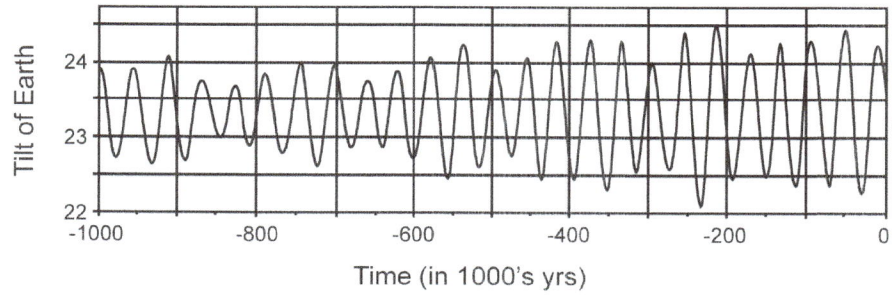

FIGURE 5.6 The obliquity cycle over the last 1,000,000 years. *(Source: Martin Williams, David Dunkerley, Patrick De Deckker, Peter Kershaw and John Chappell. 1998. Quaternary Environments. Second Edition. Arnold Press, London.)*

1. In the last 1,000,000 years, how many times has our current angle of tilt (23.5°) occurred?

2. Name another major angle of tilt that has been just as common as 23.5°. _____

3. Besides the obvious 41,000 year cycle, are there any other larger cycles apparent within the tilt cycle over the last one million years?

However, because the obliquity cycle is in constant fluctuation, the tilt of the Earth is constantly changing, thus the Tropics of Cancer and Capricorn are constantly changing. Currently the Tropic of Cancer is located at 23.44°N but it decreases every year, by roughly 13 meters. For example, the people of Taiwan built a Tropic of Cancer monument in 1908, but today, almost a century later it is roughly 1.3 km north of the true Tropic of Cancer.

If the obliquity cycle continues on its current direction and rate, how far away (in meters) from the current Tropic of Cancer will the actual Tropic be in the following years:

4. A.D. 2050 _____   6. A.D. 2500 _____

5. A.D. 2100 _____   7. A.D. 3000 _____

8. At this rate, approximately how many years would it take to reach 22°N latitude?
_____

9. As the Tropics move toward the equator, what is happening to the Arctic and Antarctic circles?

10. How could the changing tilt of the Earth affect the Earth's climates?

The Precessional Cycle

The precessional cycle states that the Earth wobbles on its axis, and because of this, the orientation of the poles are constantly changing (Figure 5.7). The periodicity of the precessional cycle is roughly 23,000 years, which means that the poles complete a full circle roughly every 23,000 years.

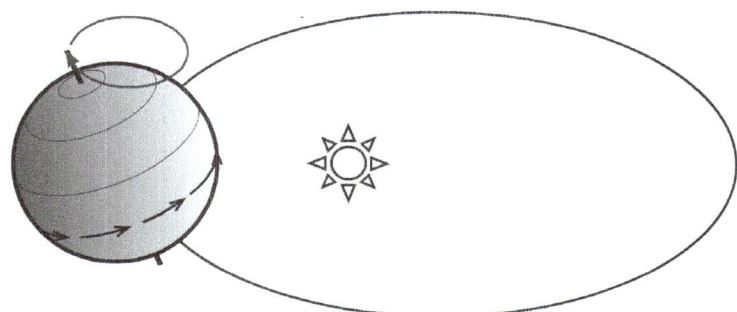

FIGURE 5.7 The Earth in its orbit around the sun, shown with the changing orientation of the poles.

In Figure 5.7, the Earth is shown tilted away from the sun, like during the month of January. This would correspond to the winter season in the Northern Hemisphere and the summer season in the Southern Hemisphere. However, in 11,500 years, or one-half of a precessional cycle, the Earth will be pointed toward the sun in the same position along its orbit. Therefore, 11,500 years from now, in the month of January we will be experiencing summer in the Northern Hemisphere and winter in the Southern Hemisphere. Essentially the precessional cycle marks the migration of the seasons.

Assuming a full cycle in 23,000 years, this means that the height of our summer gets later every year by approximately 23 minutes (or 1 day every 63 years). Currently, in the Northern Hemisphere we experience the height of summer (in terms of radiation) on June 21st. When will the height of summer be for the Northern Hemisphere in the following years:

1. A.D. 3000 _____
2. A.D. 5000 _____
3. A.D. 10,000 _____
4. A.D. 13,507 _____

Assuming this rate of change has been constant, when was the height of summer in the Northern Hemisphere for these famous historical dates:

5. Start of Roman empire (27 B.C.) _____
6. Start of the Aztec civilization (A.D. 1168) _____
7. Columbus's voyage to the Americas (A.D. 1492) _____

8. The beginning of the Holocene epoch (10,000 B.P.) _____

9. End of the last ice age (18,000 B.P.) _____

Name _____ Section _____

Atmospheric Pressure and Winds

LAB EXERCISE 6

The earth's atmosphere consists primarily of gases held to the earth by gravity. The weight of the atmosphere produces a force on the earth's surface called **pressure**. The standard unit to measure atmospheric pressure is the **millibar** (mb). Average atmospheric pressure is 1013.2 mb (equivalent to 14.7 lb/sq. inch). This is referred to as **standard sea-level pressure**, however, pressure at the earth surface varies for a given time and place. Pressure is among the most important atmospheric conditions influencing weather. It determines wind direction and speed and strongly affects the probability of precipitation. The purpose of this exercise is for you to become familiar with the relationship of pressure to winds.

Altitude and Atmospheric Pressure

Density of the atmosphere is greatest at the earth's surface (at or near sea-level) and thins with increasing height. If you travel from the earth's surface up to 5.6 km, about one-half of the total mass of the atmosphere will be below you. If you move upward another 5.6 km (11.2 km above the earth's surface), about one-fourth of the atmosphere is above with three-quarters below. Generally, regardless of the beginning elevation, the atmospheric mass decreases at a rate of 50 percent every 5.6 km increase in elevation.

1. Based on the 5.6 km rule, determine the atmospheric mass above and below for each elevation (height above sea-level) in Table 6.1. (Atmospheric pressure will be determined later.)

TABLE 6.1 DETERMINE ATMOSPHERIC MASS AND PRESSURE

Height Above Sea-Level (kilometers)	Percent Atmospheric Mass Above	Percent Atmospheric Mass Below	Atmospheric Pressure (mb)
39.2	_____	_____	_____
33.6	_____	_____	_____
28.0	_____	_____	_____
22.4	_____	_____	_____
16.8	_____	_____	_____
11.2	_____	_____	_____
5.6	_____	_____	_____
0.0 (sea-level)	100	_____	_____

2. Plot the designated values for *percent atmospheric mass* from the above table on Figure 6.1. After plotting these values, *carefully draw a smooth curving line* that best fits these data and shows the changing relationship between elevation and pressure.

   Based on the graph you have created on Figure 6.1 answer the following questions.

1. The highest elevation in your state is _____ m.

   What percentage of the atmosphere lies above this elevation? _____ percent

2. The highest mountain in North America excluding Alaska is Mt. Whitney. Its peak is 4418m above sea-level.

   What percentage of the atmosphere is above Mt. Whitney? _____ percent

3. Mt. Everest is the highest mountain in the world. Its peak is 8848m. What percentage of the atmosphere is above Mt. Everest? _____ percent

4. What is the height above which only 5 percent of the atmosphere's mass lies? _____ km

Barometric pressure is the weight of the atmosphere and can be measured at the earth's surface or at some point above the surface. With increasing elevation there is less atmospheric mass and a corresponding decrease in atmospheric pressure. Only 50 percent of the atmosphere lies above 5.6 km. Therefore, atmospheric pressure at that elevation is 50 percent of 1013.2 mb (standard sea-level pressure), or 506.6 mb.

1. On the right side of Table 6.1 is a column labeled "Atmospheric Pressure." Determine the atmospheric pressure (mb) for each elevation shown.

2. Locate and label atmospheric pressure values on Figure 6.1.

3. Use Figure 6.1 to determine barometric pressure for the following elevations:

   height above which 50 percent of the atmosphere lies: _____ mb

   height above which 5 percent of the atmosphere lies: _____ mb

   height of Mt. Everest: _____ mb

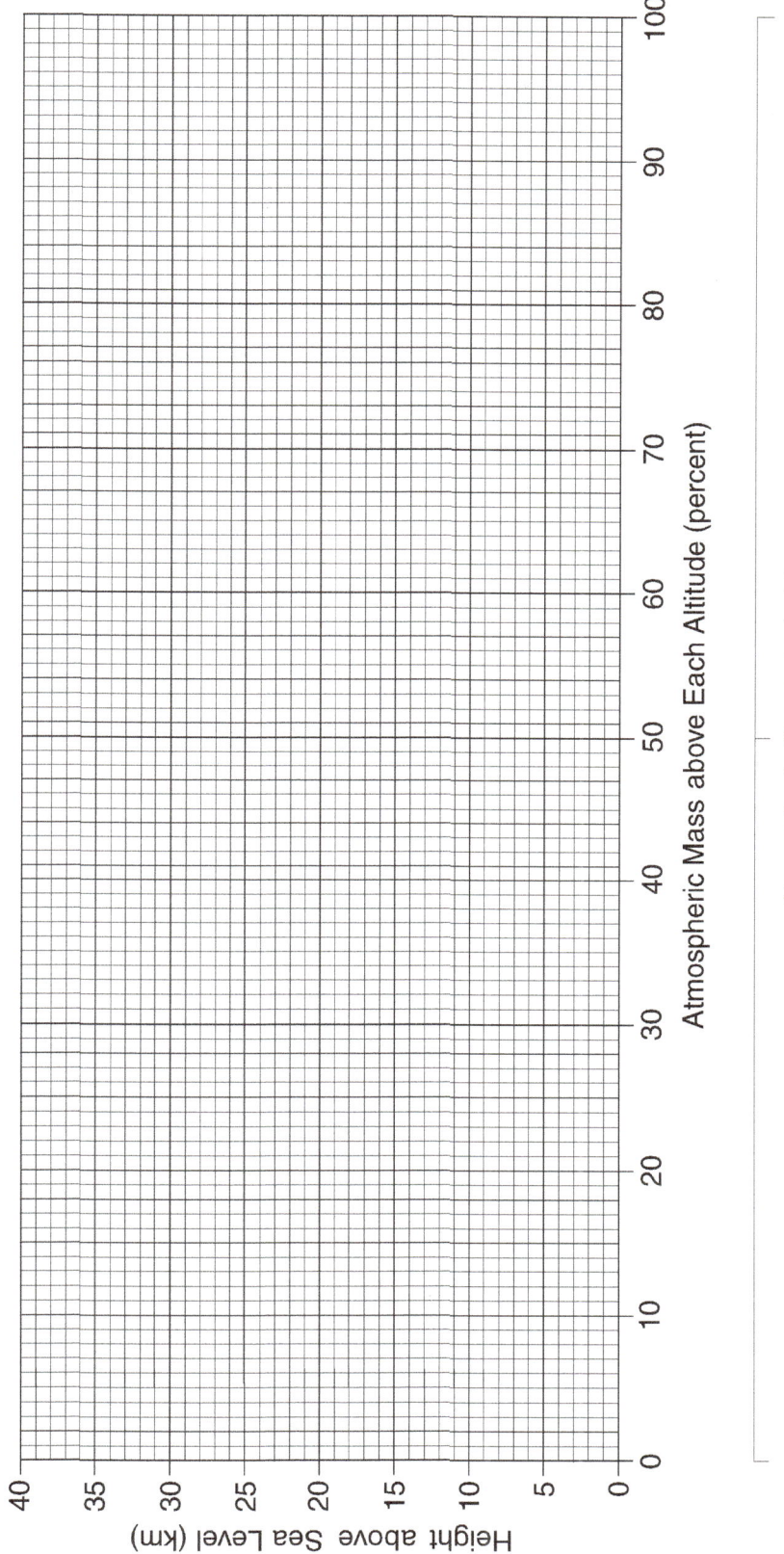

FIGURE 6.1 Determine atmospheric pressure.

Pressure and Winds

Small and large scale differences in pressure from one place to another are caused by differences in temperature at the earth's surface. The differences in pressure result in a **pressure gradient**. Pressure gradients at the earth's surface are the driving force of winds. Where there is a strong pressure gradient (a large difference in pressure over a short distance) winds will be strong. Conversely, a weak pressure gradient will result in weaker winds. **Isobars** are lines on a barometric pressure map that connect points of equal pressure. The strength of a pressure gradient can be shown on a weather map by the spacing of isobars. The closer the spacing of isobars, the steeper the pressure gradient. *The pressure gradient force is always directed perpendicular to isobars, away from high pressure and toward low pressure.*

Figure 6.2 is a small section of a surface map showing atmospheric pressure. A larger pressure map would show circular isobars indicating areas of high and low pressure. On this figure:

1. Indicate areas of highest (H) and lowest (L) pressure.

2. Use an arrow to show direction of the pressure gradient.

3. Indicate where the pressure gradient is strong (s) and weak (w), and where the winds are strongest (st).

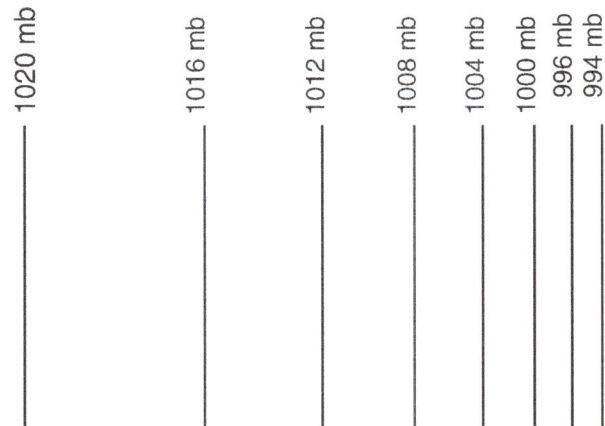

FIGURE 6.2 Section of surface map showing atmospheric pressure.

Coriolis Effect

Pressure gradient is the driving force of winds and directs winds perpendicular to isobars. But winds do not generally move perpendicular to isobars because of the Coriolis effect. **Coriolis effect** is an apparent deflection of winds caused by the earth's rotation. *Every free-moving object, including winds, deflects to the right of its path of motion in the northern hemisphere, and to the left of its path of motion in the southern hemisphere.* The movement of air near the earth's surface is at an angle across the isobars toward the area of low pressure.

Isobars on weather maps are closed circles, indicating areas of high and low pressure. Low pressure cells are called **cyclones** and show where air is rising. High pressure cells are **anticyclones**, showing where air is sinking. In the northern hemisphere, winds circulate counter-clockwise as they move toward the center of low pressure (cyclones). In the northern hemisphere, winds blow clockwise around high pressure cells (anticyclones). In the southern hemisphere, Coriolis effect deflects winds to the left (opposite direction from the northern hemisphere), so circulation is clockwise around cyclones and counter-clockwise around anticyclones.

Figure 6.3 is a simplified map of barometric pressure at the earth's surface *in the northern hemisphere.*

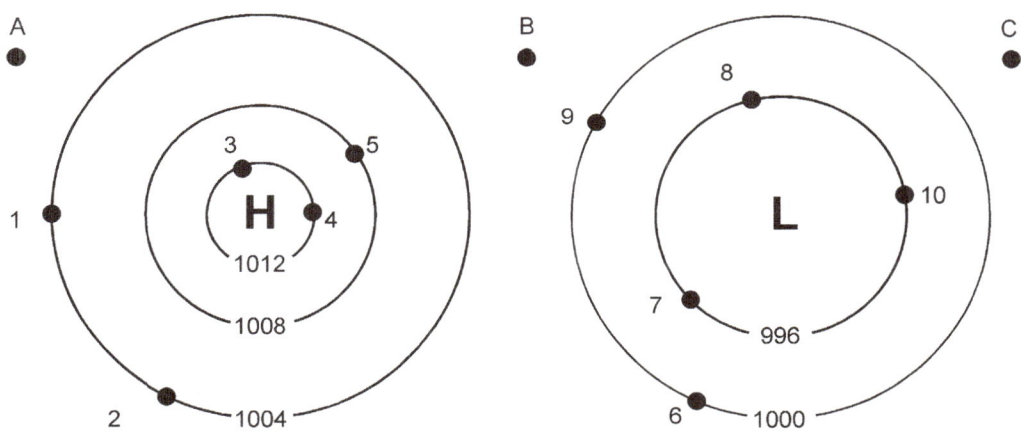

FIGURE 6.3 Northern hemisphere surface pressure map.

1. On this figure draw short straight arrows to indicate wind direction for locations 1–10.

2. The wind would blow *from* what direction at:

    Point A: _____

    Point B: _____

    Point C: _____

Drawing Isobars

Barometric pressure shown by isobars is one of the most important features of a weather map. Drawing isobars will help you to determine how surface winds move over large areas.

On Figure 6.4 is a map showing barometric pressure (mb) observed simultaneously at many National Weather Service stations. Draw isobars for the entire map using an interval of 4 mb: 992, 996, 1000, 1004, 1008, etc. Label the isobars and the center of high pressure (H) and low pressure (L). Be sure that isobars do not cross. Use a pencil and draw lightly until you are sure the isobar positions are correct, then darken the lines. Finally, draw short straight arrows across the isobars to show the directions of surface winds.

Answer the following questions based on the barometric pressure map you have created.

1. What is the highest pressure on this map and where is it located?

   highest pressure: _____ mb

   location: _____

2. What is the lowest pressure on this map and where is it located?

   lowest pressure: _____ mb

   location: _____

3. In what direction is the wind blowing *from*

   at your location: _____

   _____ : _____

   _____ : _____

   _____ : _____

A general rule about the relationship between barometric pressure and surface wind is: *Stand with the wind at your back with your arms outstretched to either side, and the lower pressure will be towards your left.* This is called Ballot's Law. Although not really a law of science, it works quite well with two provisions. First, you must be in the northern hemisphere. Second, your left arm should be aimed halfway between straight left and straight ahead.

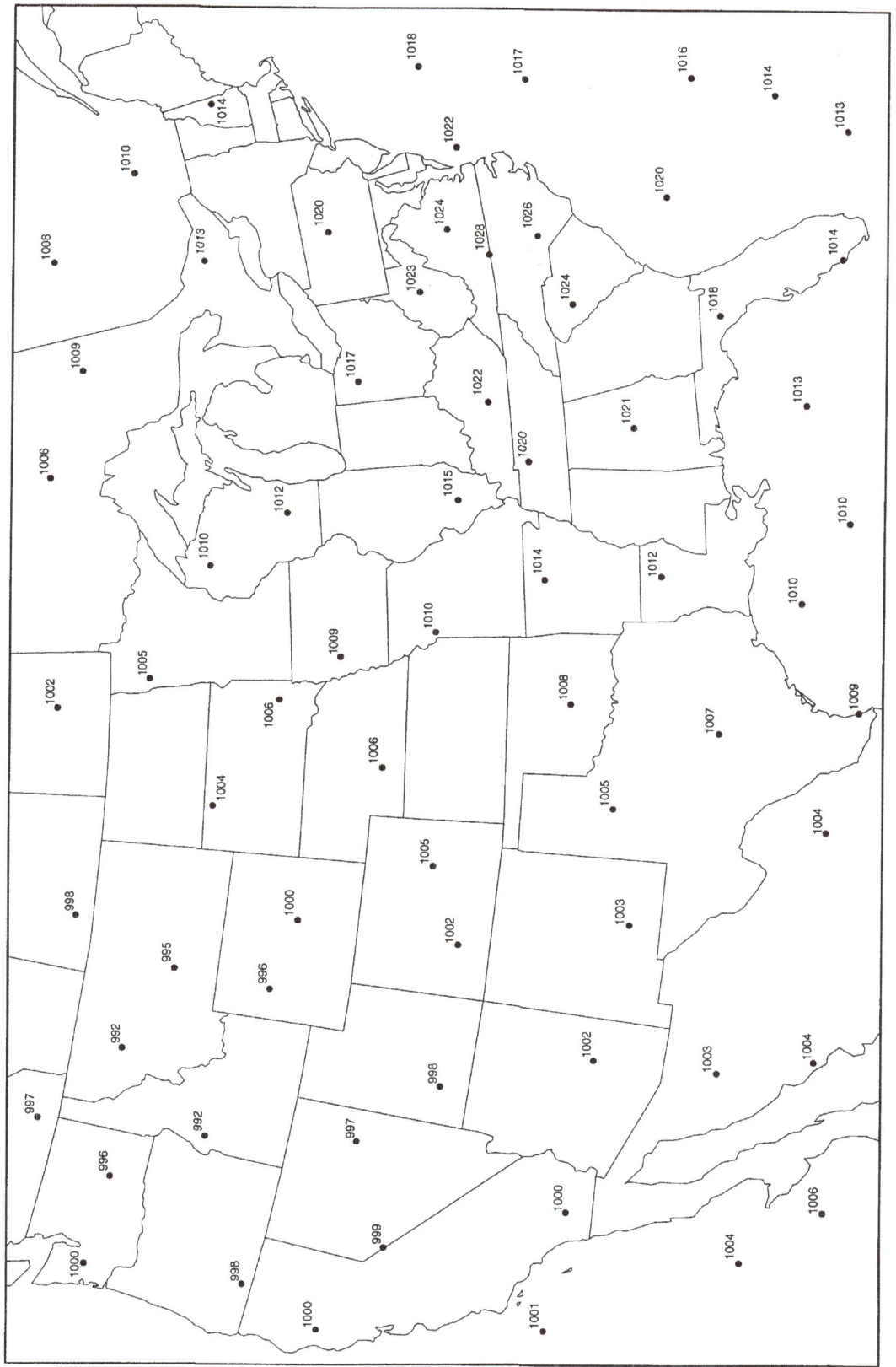

FIGURE 6.4 Barometric pressure (mb) at different National Weather Service Stations.

Use this technique to determine the position of high and low pressure relative to your position. Measure the following information in an area away from buildings or topographic features that could obstruct winds:

wind direction: _____

wind speed: _____ km/hr

barometric pressure: _____ mb

Figure 6.5 is a map of the United States. On this map (1) write the barometric pressure reading you have recorded at your location and (2) draw a short straight arrow indicating wind direction. Taking into account how surface winds circulate around cyclones (as shown in Figures 6.3 and 6.4) indicate the likely position of the center of low pressure (L). Draw isobars around the Low. Remember that surface winds move at an angle across isobars toward the center of low pressure. Also, include the possible position for high pressure (H) and draw additional isobars. You have already shown wind direction at your location. Draw a few short straight arrows across the isobars to show the directions of surface winds at other locations.

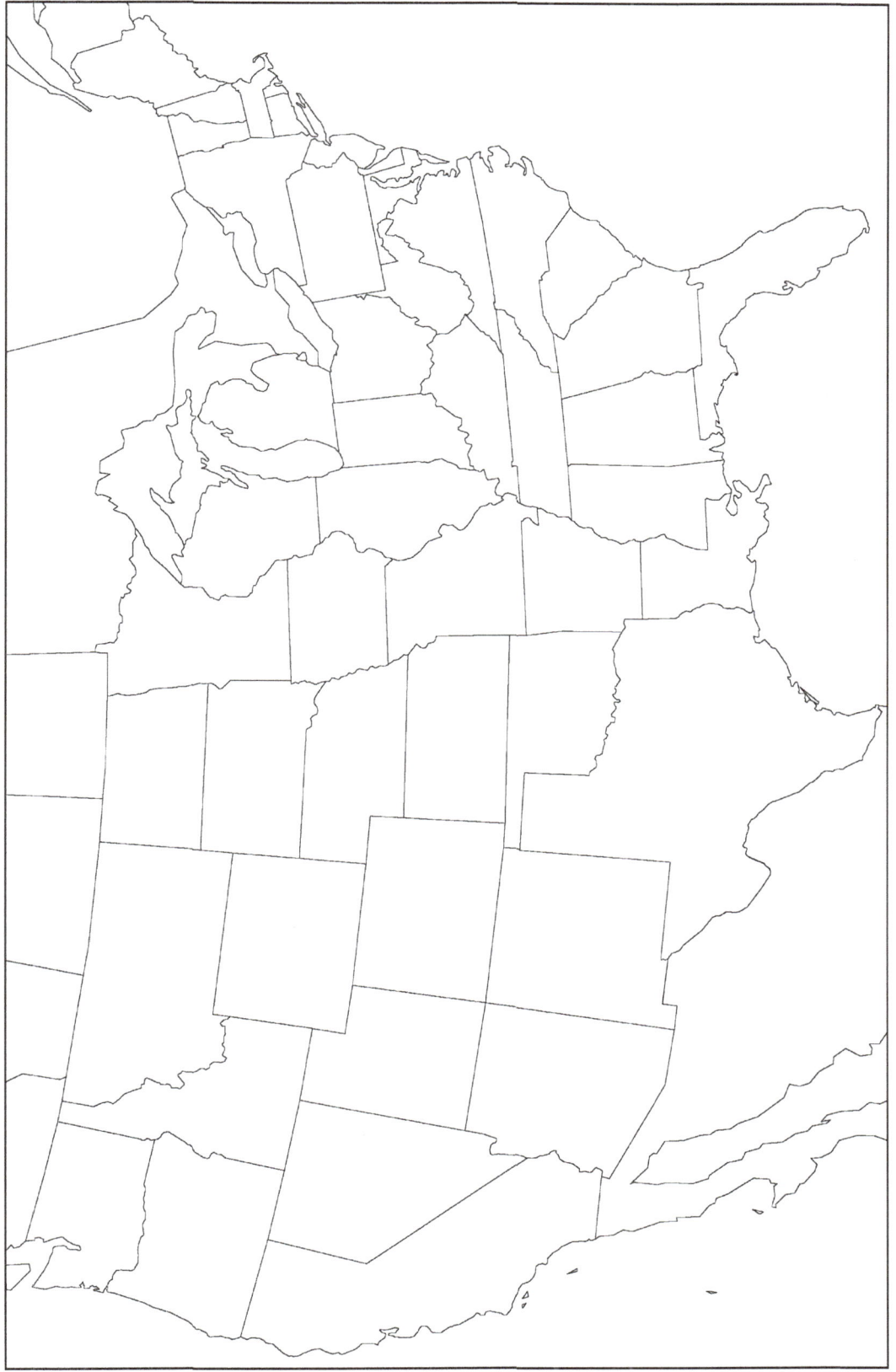

FIGURE 6.5 Map of the United States.

Name_____ Section_____

Global Atmosphere and Ocean Circulation

LAB EXERCISE 7

In the previous chapters, the basics of atmospheric motion were discussed; namely the Pressure Gradient Force and the Coriolis Force. In the reality, these forces interact with gravity and the physical characteristics of the Earth to produce persistent patterns of atmospheric circulation around the globe. Please keep in mind that these patterns of general circulation are long-term averages and that local conditions may vary from this pattern at any time. However, by studying the general circulation of the atmosphere we can gain insight into the patterns of weather and climate across the globe, which are greatly influenced by this large-scale circulation.

The physical properties of the atmosphere and the Earth combine to produce three circulation cells per hemisphere, as shown in Figure 7.1. The tropical circulation cell is known as the **Hadley Cell**. The Hadley Cell is driven by a belt of low pressure, located near the equator, called the **Inter-tropical Convergence Zone (ITCZ)**. The air at the ITCZ rises to the top of the atmosphere and moves poleward in each direction, cooling as it travels away from the equator. Though some of this air actually reaches the poles, the majority of it sinks back down to the surface of the Earth around 25–30° in either hemisphere. This belt of semi-permanent high pressure is known as the **Sub-tropical Highs**. While some of this sinking air returns to the ITCZ (thus completing the Hadley Cell), a portion of it moves poleward into the mid-latitudes. The circulation cell in the mid-latitudes is known as the **Ferrel Cell**, but it is not a true cell. Airflow in the Ferrel Cell is poleward, both at the surface and aloft and therefore does not complete a full circuit.

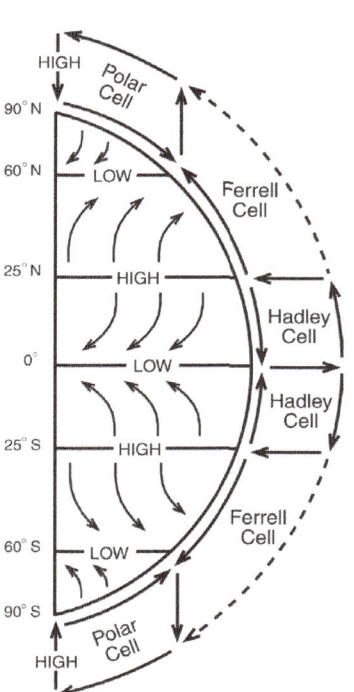

The **Polar Cell** is driven by the permanent zone of high pressure over each of the poles. Here, extremely cold and dry air sinks to the surface and moves toward the equator in both hemispheres. This air eventually converges with the midlatitude air and creates a belt of low pressure around 60° in either hemisphere known as the **Polar Front**. The warmer air from the Ferrel Cell is forced to rise and continues toward the poles, completing the Polar Cell.

FIGURE 7.1 General circulation of the atmosphere, showing both winds at the surface and aloft.

As air flows between theses belts of high and low pressure they are constantly being deflected by the Coriolis Force, to the right in the Northern Hemisphere and to the left in the Southern Hemisphere. This deflection creates persistent belts of either easterly or westerly flow within each circulation cell. Using what you have learned about the general circulation of the atmosphere, in what direction are the prevailing winds at the following locations:

1. Surface air at 10°N _____
2. Air aloft at 40°S _____
3. Air aloft at 10°N _____
4. Surface air at 75°N _____
5. Air aloft at 90°S _____
6. Surface air at 0° _____

7. In August, 1492 Christopher Columbus sailed from near Huelva, Spain (37°N) and landed in Cuba (22°N) almost 3 months later (Figure 7.2). Because this voyage occurred nearly two hundred years before the invention of the steam engine, Columbus had to rely on the prevailing winds to power his ships to the new world and back home. Using the wind belts discussed above, describe the course that Columbus needed to take (round-trip) so that he never had to sail against the prevailing winds.

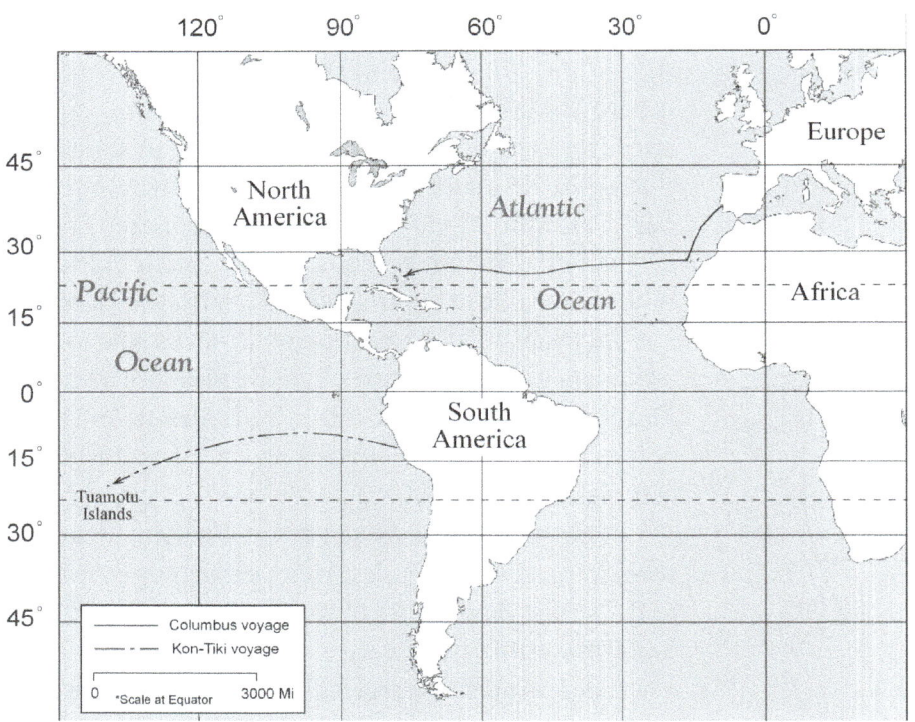

FIGURE 7.2 Columbus' first voyage, 1492–1493 and Thor Heyerdahl's Kon-Tiki voyage.

The Trade Winds

When the ITCZ is located near the equator, surface winds converging on the ITCZ from the Northern and Southern Hemisphere, are bent to the right and to the left (respectively) by the Coriolis Force. This results in easterly belt of winds within the Hadley Cell (Equator to roughly 25°) in each hemisphere. This flow is known as the Trade Winds (Figure 7.3).

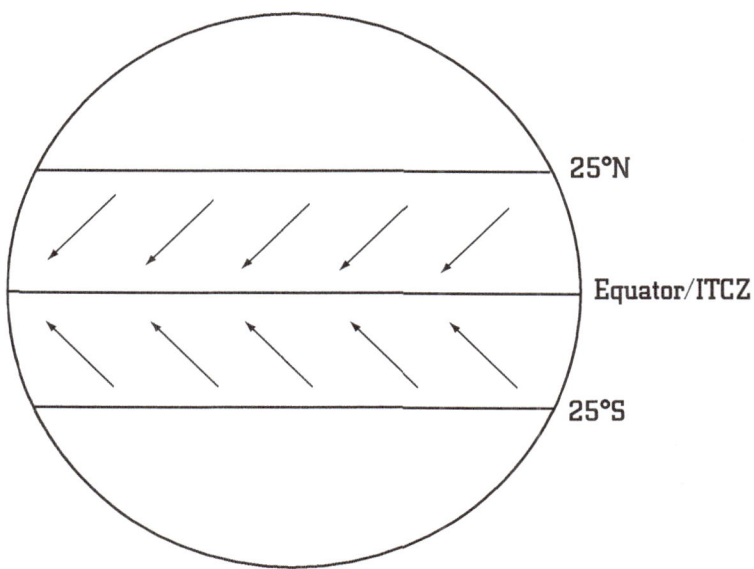

FIGURE 7.3 Surface flow within the Hadley Cell.

The trade winds are a very consistent and persistent feature, with an average wind speed of 2–4 mph. In 1947, Norwegian explorer Thor Heyerdahl sailed roughly 4,300 miles from Callao, Peru to the Tuamotu Islands, French Polynesia using only sails and a primitive balsa wood raft, which he named the Kon-Tiki (Figure 7.2). His goal was to prove that ancient South American peoples could have reached the islands of the South Pacific using only the technology available to them at the time. During his voyage he noted that, "The wind did not become absolutely still—we never experienced that throughout the voyage. . . . There was not one day on which we moved backward toward [South] America, and our smallest distance in twenty-four hours was 9 sea miles, while our average run for the voyage as a whole was 42½ sea miles in twenty-four hours."

Using Thor Heyerdahl's average speed in the trade winds during the voyage described above, and Figure 7.2, answer the following questions:

1. Between what lines of longitude did the Kon-Tiki expedition take place? _____

2. Callao, Peru is at 12°S latitude. How many degrees of latitude did the Kon-Tiki stray north and south of its starting location? _____

3. How many days was Heyerdahl at sea during the Kon-Tiki expedition? _____

4. Based on Heyerdahl's average speed, how many days would it have taken Columbus to sail from Huelva, Spain to Cuba? (Hint: refer back to Figure 7.2) _____

5. Was Columbus' average speed in the Atlantic Ocean greater or less than Heyerdahl's average speed in the Pacific? (Hint: refer back to question 6 for clues) _____

The Westerlies

The sub-tropical highs are located at roughly 25–30° in either hemisphere. Air under the influence of this high pressure is forced to sink back down to the surface. Once the air reaches the surface, some of the air is turned poleward and becomes the surface air in the mid-latitudes, between roughly 25–60°N or S latitude (refer back to Figure 7.1). As this midlatitude air flows poleward, it is turned to the right or the left in the Northern and Southern Hemispheres, respectively, by the Coriolis Force. The resulting westerly flow at the mid-latitudes in both hemispheres is known as the prevailing westerlies (Figure 7.4).

Because of the westerly flow, most of the weather in the mid-latitudes travels from west to east. For example, a weather system in eastern Texas today, will most likely be affecting Louisiana in the very short future. Please note that this westerly flow is the general atmospheric circulation within the mid-latitudes. On any given day, depending on local conditions, the wind direction and flow could be almost anything.

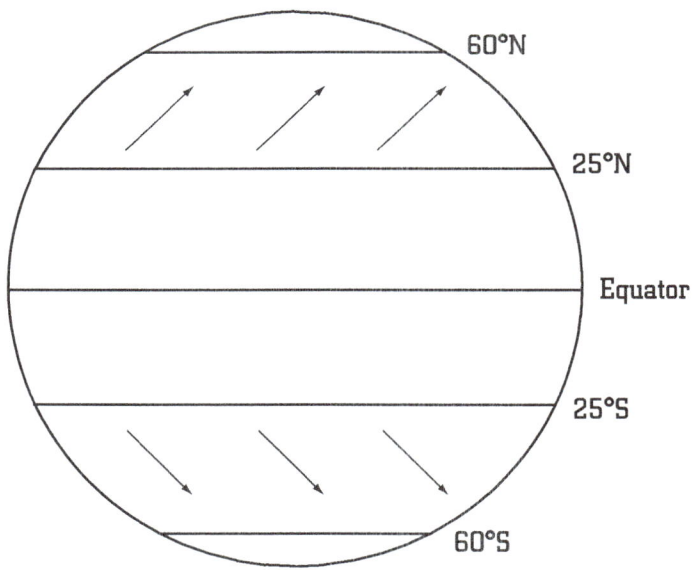

FIGURE 7.4 The prevailing westerlies.

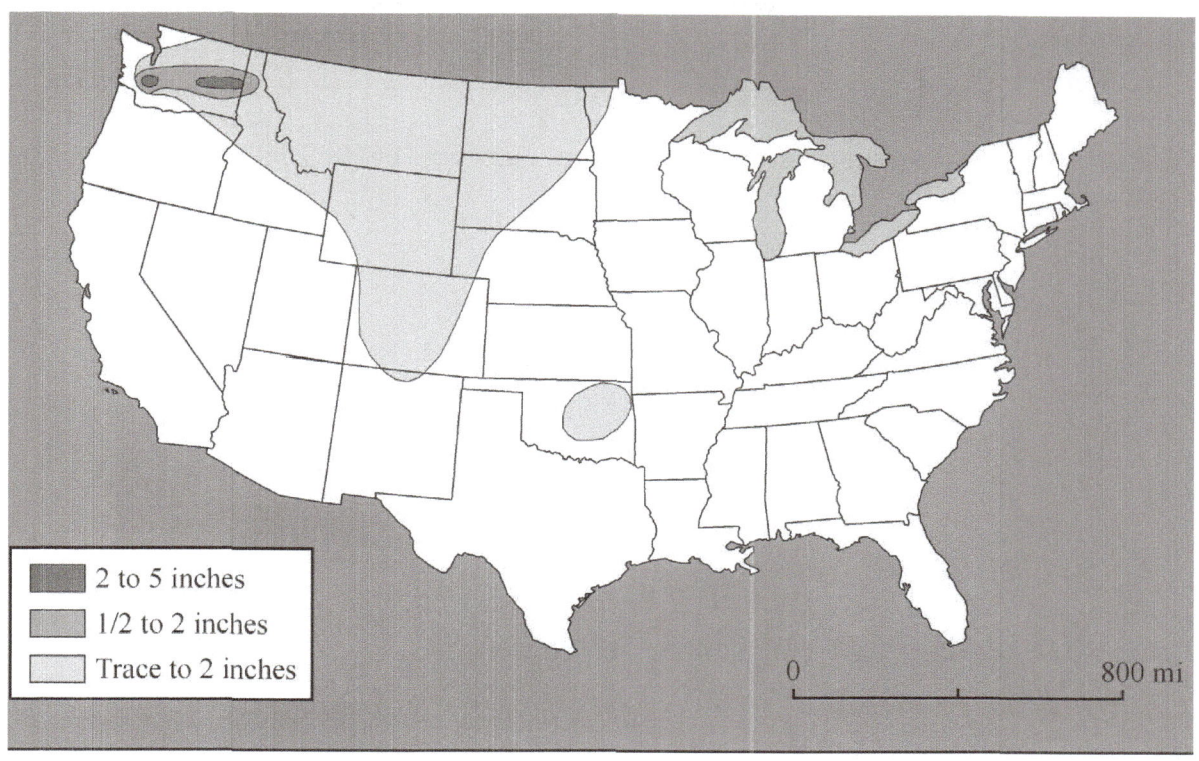

FIGURE 7.5 The distribution of ash fallout from the 1980 Mt. Saint Helens eruption.

On May 18, 1980, Mt. Saint Helens erupted in western Washington. During the 9 hour eruption, this stratovolcano discharged over 4 billion cubic yards of material in a pillar of dust, ash, and rock that extended nearly 15 miles high. Because Mt. Saint Helens is located in the mid-latitudes, the prevailing westerlies carried this airborne ash to the east where large amounts of ash were deposited from Washington to the Midwestern United States (Figure 7.5).

Use Figure 7.5 to answer the following questions.

1. How many states received at least a trace amount of ashfall from the Mt. Saint Helens eruption? _____

2. In what general direction was the ash blown?
   Compass direction _____ Azimuth _____

3. How many kilometers away from Mt. Saint Helens could trace amount of ash be found? _____

4. In what state was the southern-most extent of the ash fall? _____

   Eastern-most? _____

5. Figure 7.5 shows a disjunct distribution of the ash produced by Mt. Saint Helens. What could have caused this?

Seasonal Shifting of the Wind Belts
Counter-Trade Winds

Up to this point, we have associated the location of the ITCZ with the Equator. Though the Equator is a convenient average location for the ITCZ, it is not a static feature. In reality, the location of the ITCZ varies throughout the year, as it follows the direct rays of the sun. During the Northern Hemisphere summer (June–August), the ITCZ will typically be located north of the Equator as much as 10–15° of latitude. The variation from the Equator is much more pronounced over land, and lessened over large bodies of water. Likewise, during the Northern Hemisphere winter (December–February), the ITCZ will typically be located south of the Equator following the direct rays of the sun. Even though the ITCZ may be located north or south of the Equator, it still marks the center of the Hadley Cell circulation, and the point at which the converging winds from the Northern and Southern Hemisphere meet.

Figure 7.6 shows the ITCZ in a position north of the Equator. Winds north of the ITCZ travel south into the low pressure and are bent to the right by Coriolis producing easterly winds. Winds from the Southern Hemisphere travel north into the ITCZ, but while they are still in the Southern Hemisphere, are bent to the left by Coriolis producing easterly winds. However, once they cross the Equator (still on their way to the ITCZ) they are now bent to the right by Coriolis producing a westerly flow to the south of the ITCZ. This belt of westerly flow within the trade winds is known as the counter-trade wind.

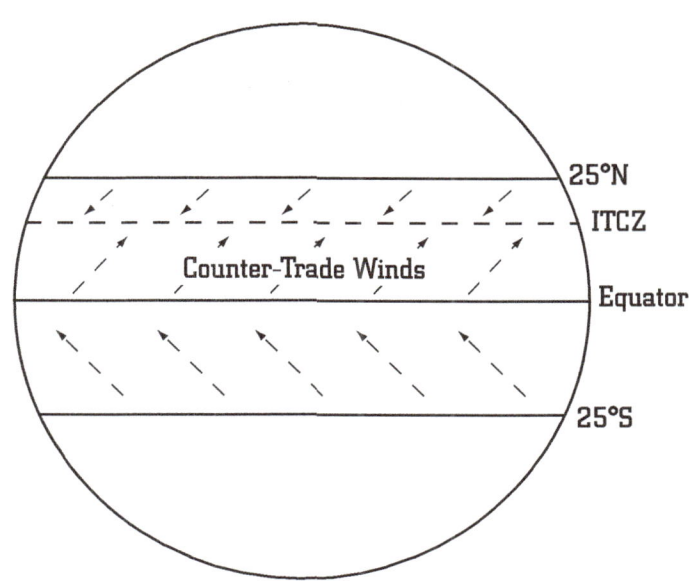

FIGURE 7.6 Surface flow when the ITCZ is located north of the equator.

Using what you've learned from this discussion on the counter-trade winds, draw in the surface flow on the following globes (Figures 7.7 and 7.8):

1.

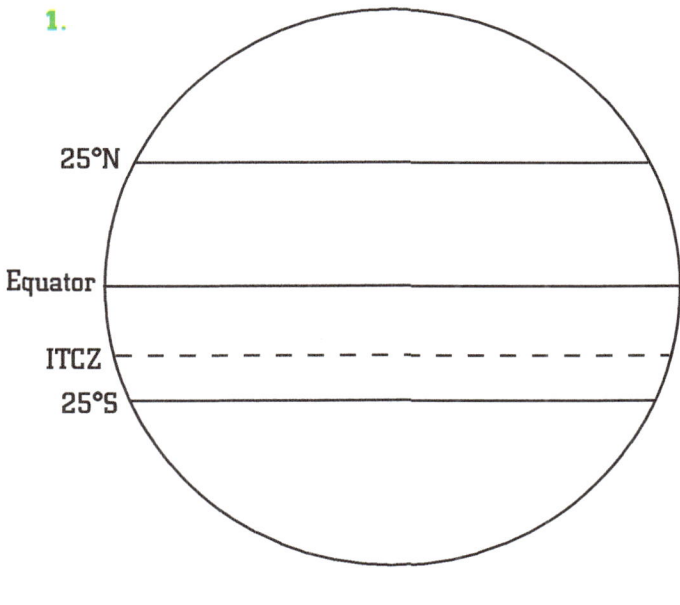

FIGURE 7.7

2.

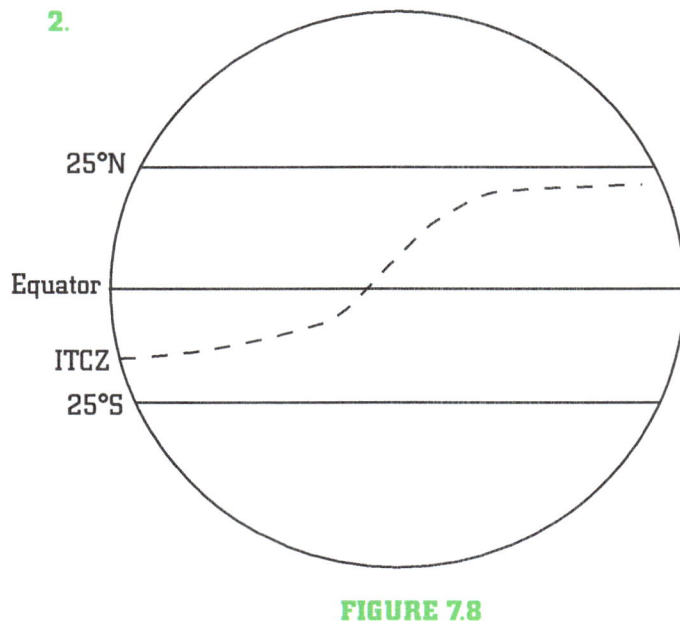

FIGURE 7.8

3. Caracas, Venezuela is coastal city that is located at roughly 10°N. The ITCZ is typically located well to the north of Caracas during the Northern Hemisphere summer, and therefore it is affected by the counter trades. During the summer months, from what direction should Caracas get their prevailing surface winds? _____ During the winter? _____

Polar Front Jet Stream

Just as the Hadley Cell shifts north or south with the direct rays of the sun, so does the Ferrel Cell and the location of the Polar Front. Remember, the Polar Front is the dividing line between the Ferrel Cell and the Polar Cell (refer back to Figure 7.1). Therefore, the polar front divides the warmer mid-latitude air from the much colder air within the Polar Cell. Because of the great temperature contrast in this area, storms are very common along the Polar Front.

In the upper atmosphere, right along the Polar Front, sits the Polar Front Jet Stream. The Jet Stream is a corridor of very fast moving winds produced by strong pressure difference. Strong differences in temperature equate to strong differences in pressure, therefore, wherever the warm mid-latitude air and the cold Polar air meet you will find the Jet Stream.

During the Northern Hemisphere summer, the Jet Stream is pushed further to the north by the shifting Ferrel Cell. Likewise, during the Northern Hemisphere winter, the Jet Stream is allowed to migrate further to the south (Figure 7.9).

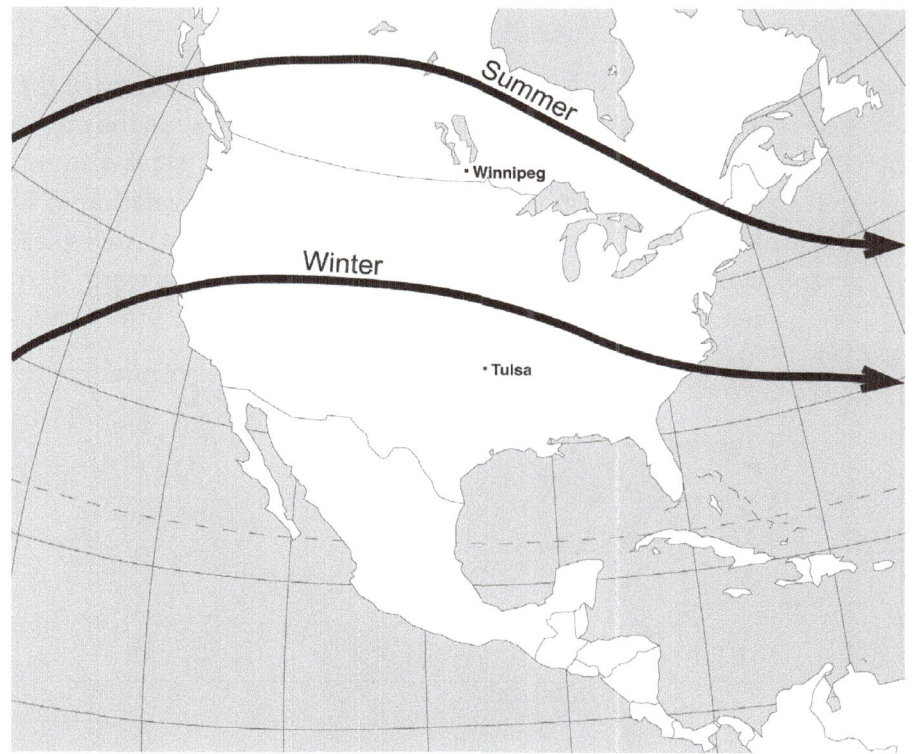

FIGURE 7.9 Map of North America showing the seasonal migration of a hypothetical jet stream.

Using the map in Figure 7.9, answer the following questions.

1. Compare and contrast the cities of Tulsa, Oklahoma, and Winnipeg, Manitoba, Canada. Explain what type of airmass (subtropical or polar) will likely dominate each city in both the winter and the summer and why?

2. Knowing that the Polar Front is an area where storm development is likely, during which season do you think storms are more prevalent in Tulsa and Winnipeg based on the location of the jet stream?

Surface Winds and Ocean Currents

As winds blow across the ocean, they drag the surface waters in the direction of their flow, producing surface ocean currents. Figure 7.10 shows the connection between the prevailing surface winds and the ocean currents in the Atlantic Basin. The result is a clockwise rotation of the ocean currents in the Northern Hemisphere and a counter-clockwise rotation in the Southern Hemisphere. Because of this circulation, the western edges of continents are typically bordered by cold ocean currents, while the eastern edges are typically bordered by warm currents.

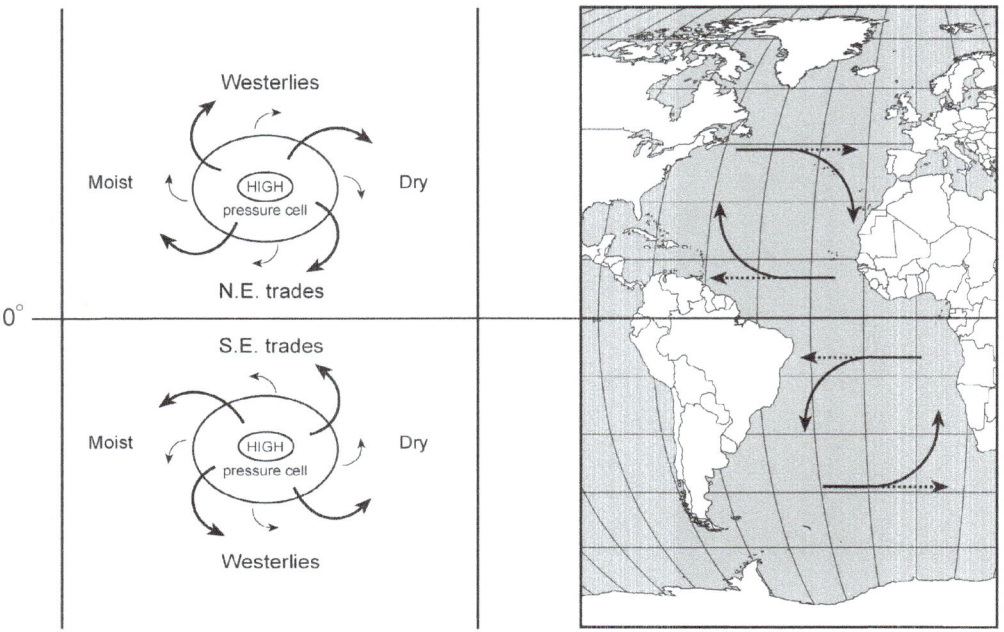

FIGURE 7.10 The connection between the prevailing surface winds and the circulation of the ocean currents in the Atlantic Basin.

1. Using the blank map of the world on page 73 (Figure 7.11), draw arrows that show the direction of the winds (wind belts) at the surface of the Earth. Next, use a red marker to draw the major warm ocean currents and their direction of circulation, and a blue marker to draw the cold. Be sure to draw the ocean circulation North and South Atlantic Ocean, and the North and South Pacific Ocean.

2. Name four countries (or states, or regions within a country) that are located along a cold ocean current:

   _____   _____   _____   _____

3. Name four countries (or states, or regions within a country) that are located along a warm ocean current:

   _____   _____   _____   _____

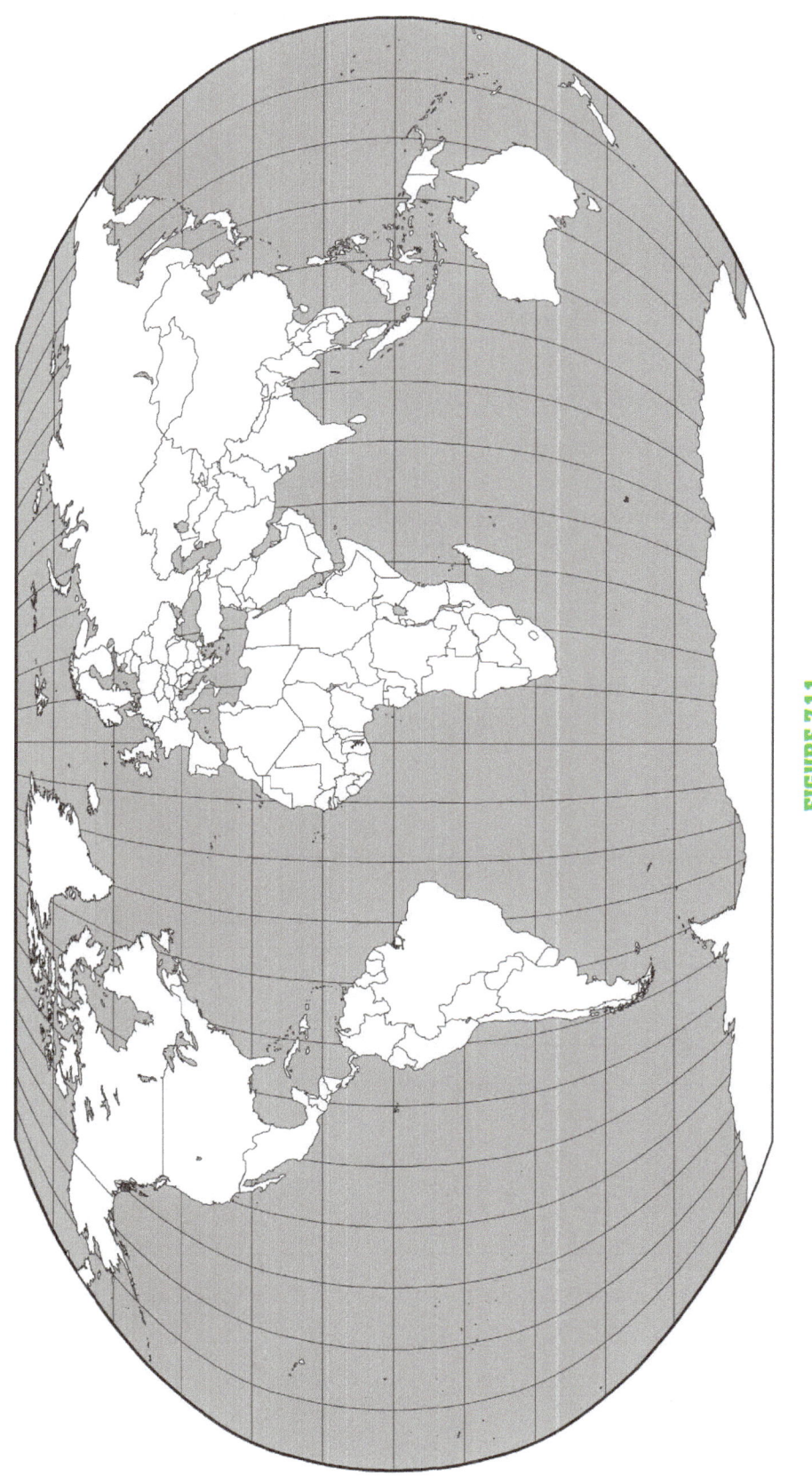

FIGURE 7.11

4. What effects would a warm and cold ocean current have on the weather and climate of the countries or regions listed on page 72?

La Nina, El Niño, and the Southern Oscillation

The trade winds produced by the Hadley Cell circulation have an enormous impact on the surface waters of the tropical Pacific Ocean. Normally, there is high pressure over the eastern Pacific (e.g., Tahiti) and low pressure over the western Pacific (e.g., Australia). The resulting winds push the surface water toward the west, which results in extra water piled up against Australia and Indonesia on the western edge of the basin. In fact, average sea-level in the western Pacific can be greater than 1.5 feet above average sea-level on the eastern edge (Figure 7.12). The removal of water from the eastern basin causes the upwelling of deep ocean water compensate for the loss. This nutrient-rich bottom water is primarily responsible for the fertile fishing grounds of the coast of South America. When this circulation is stronger and more pronounced than normal,

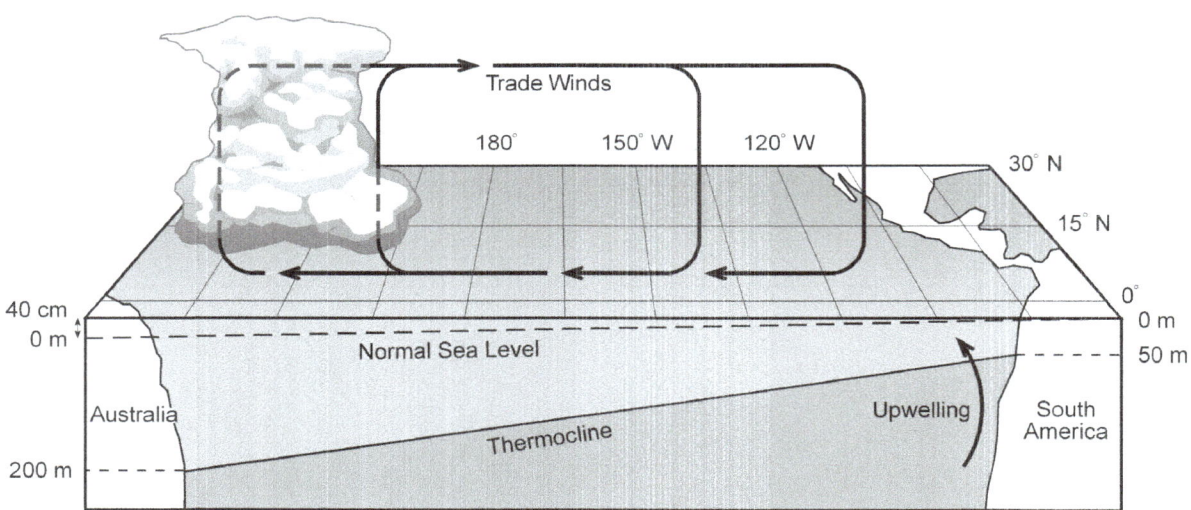

FIGURE 7.12 Diagram of normal circulation in the tropical Pacific basin. *(Adapted and corrected in R. Christopherson, Geosystems: An introduction to Physical Geography, Fifth Edition (Prentice-Hall; 2005) from C.S. Ramage, "El Nino," Scientific American, 1986.)*

the term La Nina is given to the event.

However, for reasons that are not fully understood, every 3 to 7 years there is a break-down in this Hadley Cell circulation and the trade-winds weaken dramatically or even reverse their flow. This shift in atmospheric circulation is called the Southern Oscillation. When the Southern Oscillation occurs there is usually higher pressure over Australia and Indonesia and lower pressure over Tahiti and the eastern Pacific. As a result of this wind reversal, the surface water that was piled up on the western edge of the Pacific is allowed to slosh back to the eastern Pacific. This movement of the surface waters back toward the east is known as El Niño (Figure 7.13). This event stops the upwelling of deep water off the coast of South America.

One of the ways that researchers keep track of El Niño/Southern Oscillation (or ENSO) is with the Southern Oscillation Index (SOI). The SOI, as calculated by the Climate Research Unit at East Anglia University, is the normalized difference in pressure between Tahiti and Darwin, Australia (Tahiti—Darwin). During normal conditions, the pressure over Tahiti is typically higher than Darwin, so the SOI results in a positive number. However, during an ENSO event, the pressure pattern reverses and the SOI results in a negative number. The more negative the

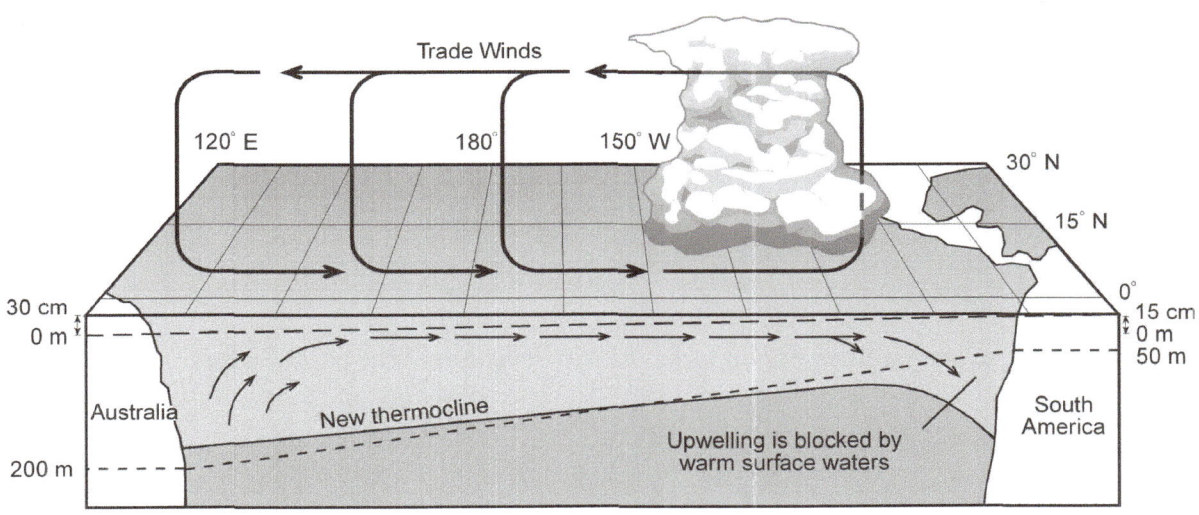

FIGURE 7.13 The circulation associated with the Southern Oscillation and the resulting flow of the tropical Pacific waters (El Niño). *(Adapted and corrected in R. Christopherson,* Geosystems: An introduction to Physical Geography, *Fifth Edition (Prentice-Hall; 2005) from C.S. Ramage, "El Nino," Scientific American, 1986.)*

Southern Oscillation Indicies

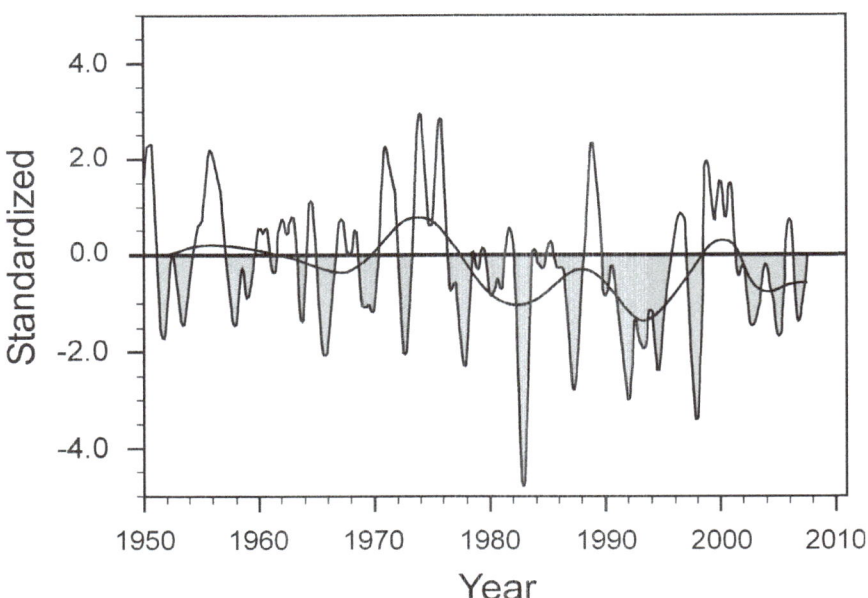

FIGURE 7.14 The Southern Oscillation Index, A.D. 1950–2007.

number, the more intense the ENSO event. Very positive numbers refer to La Nina events.

1. According to Figure 7.14, over the past 57 years how many intense El Niño and La Nina events have there been? (values over +2 or −2)

   El Niño _____   La Nina _____

2. What is the average return period for an El Niño event over the past 57 years?

   _____

3. In what years were the three most intense ENSO and La Nina events?

   ENSO _____

   La Nina _____

4. Are some decades more prone to ENSO or La Nina events? Describe any decadal, or larger-scale patterns within the SOI record.

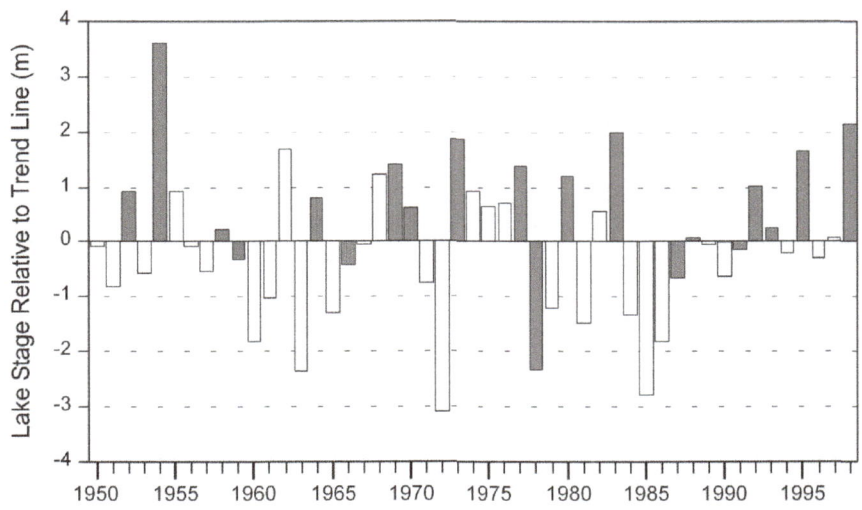

FIGURE 7.15 Annual maximum flood level on the lower Yangtze River, China, 1950–1998. Grey bars indicate El Niño years. *(Source: David Shankman, Barry D. Keim, and Song Jie. 2006. Flood frequency in China's Poyang Lake region: trends and teleconnections.* International Journal of Climatology *26: 1255–1266.)*

Though ENSO events occur in the tropical Pacific, they have world-wide effects on weather, some positive and some negative. As a case study, we will examine the effects of ENSO on the water levels of the lower Yangtze River in eastern China. This river is vitally important to agricultural production in China. The farming that occurs within the Yangtze basin accounts for nearly half of China's total crop production. Drought and floods along the Yangtze can have devastating impacts to the economy of the region.

Figure 7.15 shows annual maximum flood levels above or below 1950–1998. The white bars on the figure indicate La Nina or normal years, while the grey bars indicate El Niño events.

Based on the information in Figure 7.15, answer the following questions:

5. Describe the relationship between El Niño/La Nina years and annual floods along the lower Yangtze River?

6. Of the top 10 highest flood years, how many are El Niño years? _____

7. Of the top 10 lowest flood years, how many are La Nina years? _____

8. What challenges and/or benefits might farmers in the Yangtze River Basin face during a strong El Niño or a La Nina event?

9. Are the highest and lowest flood years evenly dispersed throughout the 48 year period, or is there some decadal variability to the data? If so, discuss and identify which decades were more prone to high water events and which were prone to low water events.

Name _____ Section _____

LAB EXERCISE

8

Atmospheric Humidity

Equipment

- psychrometer

Water vapor is a small proportion of the atmosphere, ranging from about 1–3 percent in most regions of the world. However, its abundance is one of the most important atmospheric conditions determining weather and influencing regional climates. In this exercise you will measure relative humidity and address problems that deal with the relationship between temperature, water vapor capacity, relative humidity, and condensation.

Water at the earth's surface or in the lower atmosphere occurs as liquid, solid, or gas depending primarily on temperature. Water vapor, an invisible gas, enters the atmosphere by evaporation. The amount (or density) of water vapor is referred to as **specific humidity**, and is usually expressed as grams per kilogram.

The amount of water vapor the air can hold depends on air temperature. As temperature increases, the air can hold more water. As temperature decreases, its ability to hold moisture decreases. The maximum amount of water vapor it can hold at any given temperature is referred to as **capacity**. The ratio between the **specific humidity**, the actual amount of water in the air, and **capacity** (how much it can hold at that temperature), is called **relative humidity** (RH), which is expressed as a percentage:

$$RH = (SH/CAP) * 100$$

where: SH = specific humidity
CAP = capacity

As the air temperature changes, the air's capacity to hold water and relative humidity also changes. For example, on the first line of Table 8.1 shown below, a parcel of air has a temperature of 25°C. Capacity at this temperature (as indicated in Figure 8.1) is 20 g/kg. Assume the specific humidity is 5 g/kg (although it could be higher or lower). Since the specific humidity (SH) is ¼ of capacity, the relative humidity (RH) is 25 percent. In other words, the air has ¼ of the amount of water it is capable of holding.

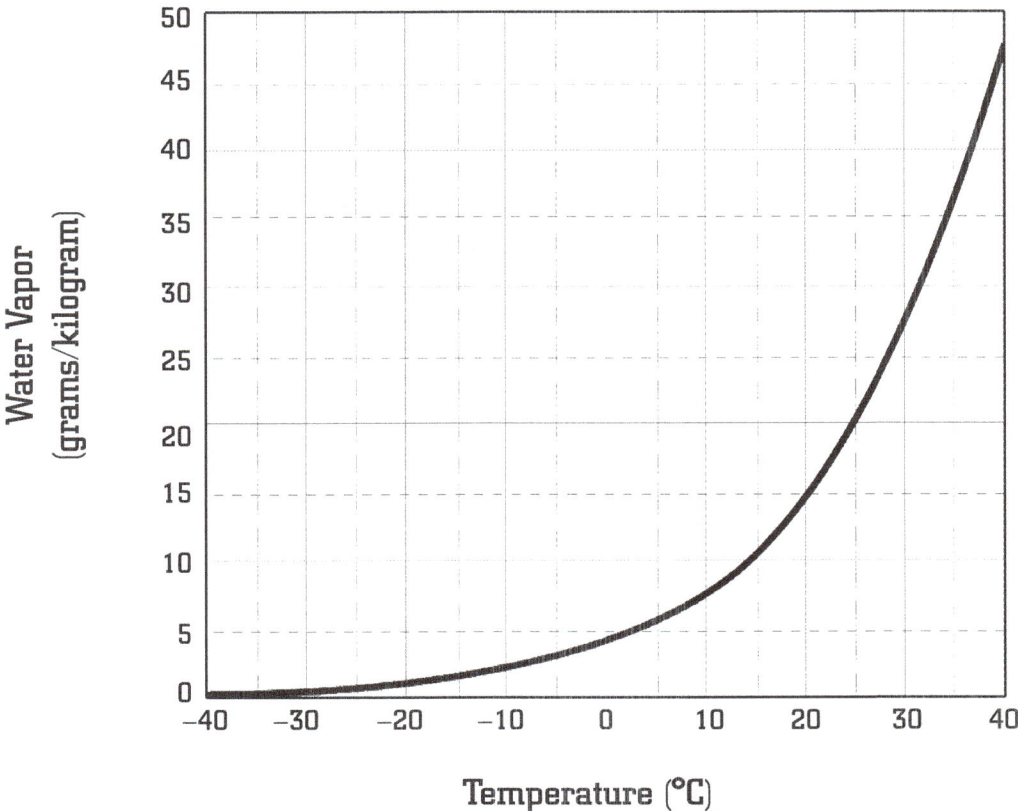

FIGURE 8.1 Saturation curve showing the maximum amount of water vapor (capacity) that can be held at a given temperature.

Now assume that the specific humidity (the amount of water in the air) does not change, but the temperature decreases. With lower temperature, the air is capable of holding less water. So, while the amount of water in the air remains the same, the capacity is reduced, and therefore the relative humidity increases. On the second line of the table shown below the air temperature decreased from 25°C to 20°C. The capacity (CAP) decreased (15 g/kg) so the relative humidity (SH/CAP) increased. As the air temperature continues to decrease, the relative humidity increases.

TABLE 8.1   DETERMINE RELATIVE HUMIDITY OF THE AIR AT VARIOUS TEMPERATURES

Temp(°C)	CAP	SH	RH(%)
	g/kg		
25	20	5	25
20	15	5	33
15	10	5	50
10	8	5	63
5	5	5	100

For the following questions refer to Figure 8.1.

1. What is the capacity of a parcel of air that has a temperature of 20°C?

   _____ grams per kilogram

2. What is the capacity of a parcel of air that has a temperature of 10°C?

   _____ grams per kilogram

3. If a parcel of air is 10°C and the specific humidity is 5 grams of water per kilogram, what is the relative humidity?

   _____ %

Fill in the blanks on the following table. Use Figure 8.1 to determine the capacity for the corresponding air temperatures.

Temp(°C)	Capacity	Specific Humidity g/kg	Relative Humidity %
30	28	10	
20		10	
15		10	
10		5	

4. Minimum daily temperature usually occurs at sunrise. Typically, the temperature rises until reaching a maximum during the mid-afternoon and then slowly declines. Based on these daily temperature changes, how would the relative humidity change during the day?

Dew point is the temperature at which the air must be cooled to reach a relative humidity of 100 percent. At a relative humidity of 100 percent the air is saturated. In Table 8.1, since the specific humidity (SH) is 5 g/kg, the air had to be cooled so that the capacity (CAP) equaled the same.

For the following questions refer to Figure 8.1.

1. What is the dew point if the specific humidity is 15 g/kg?

    _____ °C

2. What is the dew point if the specific humidity is 25 g/kg?

    _____ °C

3. If the air temperature is 10°C and the relative humidity is 50 percent, what is the dew point? (Hint: First, determine the specific humidity.)

    _____ °C

4. If the air temperature is 15°C and the relative humidity is 33 percent, what is the dew point?

    _____ °C

You can easily determine the relative humidity using a sling psychrometer, which has two thermometers. One thermometer (the dry-bulb) measures the air temperature. The other (the wet-bulb) has a cloth wick at the base which is wetted. Saturate the wick (the wet-bulb) and then swing the psychrometer for 20 seconds. Read the temperature of the wet bulb thermometer and then repeat this process until the wet-bulb temperatures do not change between readings. Record the final wet-bulb and dry-bulb temperatures. The difference between these readings is the wet-bulb depression.

The amount of evaporation from the wet bulb indicates the amount of water vapor that is in the air. If the air is near saturation little water will evaporate from the wet bulb, and therefore the wet bulb depression, or difference in temperature between the two thermometers will be small. In contrast, if the air is very dry, water will rapidly be lost from the wet bulb, and the wet bulb depression will be high.

The temperature and, therefore, humidity readings will vary on different surfaces. Vegetated surfaces and those that are shaded will typically have cooler temperatures compared to paved surfaces and those in direct sunlight. Take readings from the sling psychrometer over the different surfaces listed on the following page and fill in the following table. If you are using Celsius temperatures, refer to Table 8.2 to determine the relative humidity. Use Table 8.3 if you are using Fahrenheit.

Caution! Sling psychrometers will easily break. Please handle them carefully and make sure they do not strike other objects.

	Black Asphalt	Vegetated Surface		
		Full Sun	Shade	
Wet Bulb	_____	_____	_____	_____
Dry Bulb	_____	_____	_____	_____
Wet Bulb Depression	_____	_____	_____	_____
Relative Humidity (%)	_____	_____	_____	_____
Dew Point	_____	_____	_____	_____

1. Over which surface was the air temperature highest? Why?

2. Which surfaces had the highest relative humidity? Why?

3. Would you expect air temperature and relative humidity to be significantly different on a cloudy day? How would these conditions vary?

TABLE 8.2   PERCENT RELATIVE HUMIDITY, TEMPERATURE IN °CELSIUS

Air Temperature	\multicolumn{15}{c}{Wet Bulb Depression}

Air Temperature	1	2	3	4	5	6	7	8	9	10	11	12	13	14	15
−5	76	52	29	7											
−4	77	55	33	12											
−3	78	57	37	17											
−2	79	60	40	22											
−1	81	62	43	26	8										
0	81	64	46	29	13										
1	83	66	49	33	17										
2	84	68	52	37	22	7									
3	84	70	55	40	26	12									
4	85	71	57	43	29	16									
5	86	72	58	45	33	20	7								
6	86	73	60	48	35	24	11								
7	87	74	62	50	38	26	15								
8	87	75	63	51	40	29	19	8							
9	88	76	64	53	42	32	22	12							
10	88	77	66	55	44	34	24	15	6						
11	89	78	67	56	46	36	27	18	9						
12	89	78	68	58	48	39	29	21	12						
13	89	79	69	59	50	41	32	23	15						
14	90	79	70	60	51	42	34	26	18	8					
15	90	80	71	61	53	44	36	27	20	13					
16	90	81	71	63	54	46	38	30	23	15	8				
17	90	81	72	64	55	47	40	32	25	18	11				
18	91	82	73	65	57	49	41	34	27	20	14				
19	91	82	74	65	58	50	43	36	29	22	16	10			
20	91	83	74	66	59	51	44	37	31	24	18	12	6		
21	91	83	75	67	60	53	46	39	32	26	20	14	9		
22	92	83	76	68	61	54	47	40	34	28	22	17	11	6	
23	92	84	76	69	62	55	48	42	36	30	24	19	13	8	
24	92	84	77	69	62	56	49	43	37	31	26	20	15	10	5
25	92	84	77	70	63	57	50	44	39	33	28	22	17	12	8
26	92	85	78	71	64	58	51	46	40	34	29	24	19	14	10
27	92	85	78	71	65	58	52	47	41	36	31	26	21	16	12
28	93	85	78	72	65	59	53	48	42	37	32	27	22	18	13
29	93	86	79	72	66	60	54	49	43	38	33	28	24	19	15
30	93	86	79	73	67	61	55	50	44	39	35	30	25	21	17
31	93	86	80	73	67	61	56	51	45	40	36	31	27	22	18
32	93	86	80	73	67	61	56	51	45	41	37	32	28	24	20
33	93	86	80	74	68	63	57	52	47	42	38	33	29	25	21
34	93	87	81	75	69	63	58	53	48	43	39	35	30	26	23

TABLE 8.3 PERCENT RELATIVE HUMIDITY, TEMPERATURE IN °FAHRENHEIT

Air Temperature	\multicolumn{14}{c}{Wet Bulb Depression}
	1	2	4	6	8	10	12	14	16	18	20	22	24	26	28
24	84	73	47	22											
26	87	75	51	27	4										
28	88	76	54	32	10										
30	89	78	56	36	16										
32	89	79	59	39	20	2									
34	90	81	62	43	25	8									
36	91	82	64	46	29	13									
38	91	83	66	50	33	17	3								
40	92	83	68	52	37	22	7								
42	92	85	69	55	40	26	12								
44	93	85	71	56	43	30	16	4							
46	93	86	72	58	45	32	20	8							
48	93	86	73	60	47	35	23	12	1						
50	93	87	74	61	49	38	27	16	5						
52	94	87	75	63	51	40	29	19	9						
54	94	88	76	64	53	42	32	22	12	3					
56	94	88	75	65	55	44	34	25	16	7					
58	94	88	77	66	56	46	37	27	18	10	1				
60	94	89	78	68	58	48	39	30	21	13	5				
62	94	89	79	69	59	50	41	32	24	16	8				
64	95	90	79	70	60	51	43	34	26	18	11	1			
66	95	90	80	71	61	53	44	36	29	21	14	5			
68	95	90	80	71	62	54	46	38	31	23	16	7			
70	95	90	81	72	64	55	48	40	33	25	19	11	2		
72	95	91	82	73	65	57	49	42	34	28	21	14	5		
74	95	91	82	74	65	58	50	43	36	29	23	19	14	7	
76	96	91	82	74	66	59	51	44	38	31	25	21	16	11	
78	96	91	83	75	67	60	52	46	39	33	27	22	17	12	5
80	96	91	83	75	68	61	54	47	41	35	29	23	18	12	7
82	96	92	84	76	69	61	55	48	42	36	30	25	20	14	10
84	96	92	84	76	69	62	56	49	43	37	32	26	21	16	12
86	96	92	84	77	70	63	57	50	44	39	33	28	23	18	14
88	96	92	85	77	70	64	57	51	46	40	35	30	25	20	15
90	96	92	85	78	71	65	58	52	47	41	36	31	26	22	17
92	96	92	85	78	72	65	59	53	48	42	37	32	28	23	19
94	96	93	85	79	72	66	60	54	49	43	38	33	29	24	20
96	96	93	86	79	73	66	61	55	50	44	39	35	30	26	22
98	96	93	86	79	73	67	61	56	50	45	40	36	32	27	23

Name_____ Section_____

Adiabatic Processes and Precipitation

LAB EXERCISE
9

Equipment

- psychrometer

The rising of air is the most common cooling mechanism in the atmosphere and is the direct cause of most condensation and precipitation. The purpose of this exercise is to develop an understanding of the relationships between the rising of air, changes in relative humidity, and the elevation or level of condensation and cloud formation.

If an air mass rises, it will expand because of decreasing atmospheric pressure. As the air expands there are fewer molecular collisions that generate heat and therefore the air will cool. Rising air may cool to its dew point and clouds will form. If additional rising occurs after the dew point is reached, precipitation will likely occur. Conversely, if the air descends in altitude it is compressed and will heat, therefore, increasing its ability to hold moisture. Cloud development and precipitation will not occur if air is sinking.

The change in temperature caused by air rising or sinking is called **adiabatic cooling and heating**. It is important to make a distinction between the adiabatic process and the increase in air temperature with altitude (in an air mass that is not rising or sinking) which is referred to as the environmental lapse rate (ELR).

If air rises but has not cooled to its dew point, the air will cool at a constant rate of 10°C/1 km (1°C/100m). This is called the **dry adiabatic rate (DAR)**. If the air rises enough so that it cools to the dew point and condensation occurs, the air will continue to cool but at a lower rate of about 6°C/1 km (0.6°C/100m). This is known as the **saturated adiabatic rate (SAR)**. When air begins to descend, the temperature will heat at the dry adiabatic rate (10°C/1 km) and immediately exceed the dew point.

Orographic Precipitation

Orographic precipitation results from air cooling as it rises over mountains or topographic barriers. As it rises, the air cools at the dry adiabatic rate (10°C/1 km or 1°C/100m). As the air sinks on the leeward side, it warms at the same rate.

1. Figure 9.1 shows an air mass that is forced to rise over a mountain that is 4000m elevation. The air cools as it rises. But in this problem, assume that the air does not cool enough to

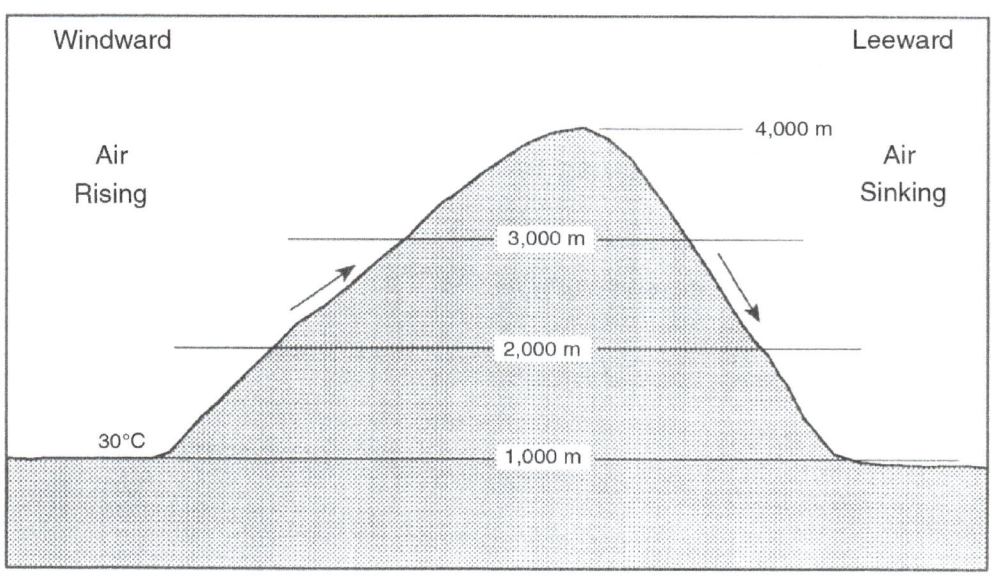

FIGURE 9.1 Calculate temperature for various elevations.

reach the dew point, so no condensation occurs. Because the relative humidity is less than 100 percent, use the DAR to determine the rate of temperature change. The temperature at 1000m elevation on the windward side of the mountain is 30°C. Calculate and indicate the temperature for the elevations shown on Figure 9.1.

If your calculations are correct, temperature should be the same on both the windward and leeward sides of the mountain at the same elevation.

2. Figure 9.2 shows an air mass that is forced to rise over a mountain at another location. The mountain summit is 4000m. The temperature at the foot of the mountain (1000m elevation) is the same as in the previous problem. But in this case assume that the humidity is higher so that condensation occurs at 2000m as the air ascends and cools on the windward side.
    Remember, use the dry adiabatic rate if the relative humidity is less than 100 percent. Use the saturated adiabatic rate only if air continues to rise after the dew point is reached.
    Calculate and indicate the temperature for the elevations shown on Figure 9.2.

After reaching the summit the air begins to descend and is adiabatically heated. The warmer air has a higher water vapor capacity than at the summit and therefore the relative humidity immediately drops below 100 percent.
   a. If your calculations are correct, air temperature at the same elevation on the windward and leeward sides of the mountain are different. Why?

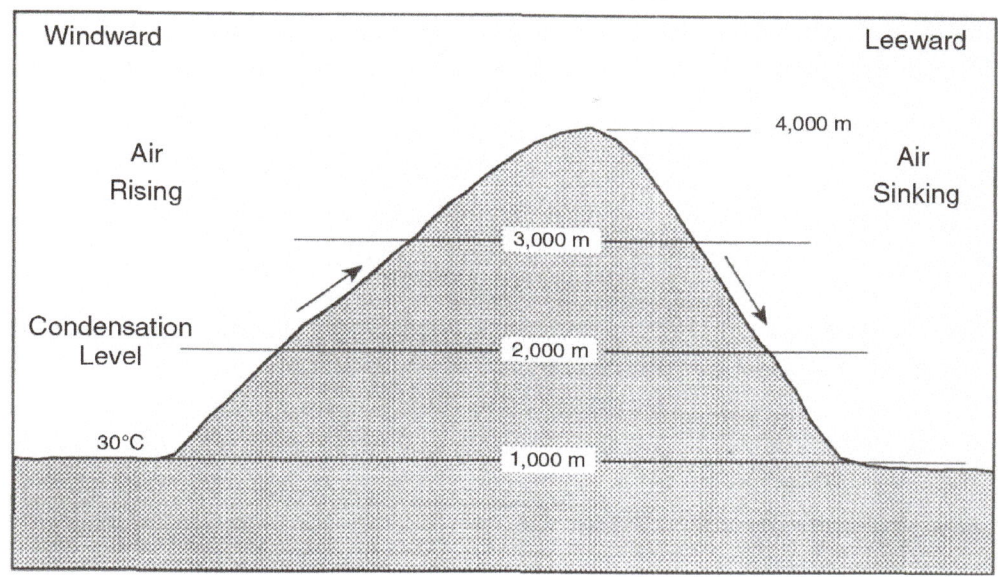

FIGURE 9.2 Calculate temperature for various elevations.

b. The relative humidity is usually lower on the leeward side of mountains. What are the two factors that cause drier conditions there?

(1) _____

(2) _____

c. A **rainshadow** refers to the dry conditions on the leeward side of mountains. This is one of the major causes of desert regions in the world. Identify a rainshadow region in North America and the mountain range(s) causing this effect.

Convection

The ascent of air is not always forced as demonstrated in the previous problems. **Convection** refers to air that rises spontaneously or freely. The earth's surface heats unevenly and some surfaces have a higher temperature than others at the same time. Since the earth's surface is the major heat source for the lower atmosphere, air temperature near the ground will vary from one place to another.

If a parcel of air is warmer than the surrounding air, it will tend to rise, cooling at the dry adiabatic rate. Even though the air is cooling as it rises, as long as it remains warmer than the surrounding air it will continue to rise. In most cases, the air cools and eventually sinks back toward the surface before reaching the dew point, and therefore, no condensation occurs.

However, if a parcel of air rises high enough to reach the dew point, clouds will form.

In the following problems, use the DAR to calculate the temperature of a rising parcel of air, and the Environmental Lapse Rate (ELR) to calculate the temperature of the surrounding air.

Environmental Lapse Rate (ELR) = 7°C/1000m
(for the surrounding air that is not rising)

Dry Adiabatic Rate (DAR) = 10°C/1000m

As shown in Exercise 8, the dew point of a parcel of air can easily be calculated at the surface. However, the dew point changes as air rises. Remember that absolute humidity is a measure of the weight of water vapor per volume of air (g/m^3). As air rises it expands. Because of the change in volume, as air rises the dew point decreases. To accurately determine the level of condensation, you must take into account the **dew point lapse rate (DPLR)**, which is a decrease in the dew point of 2°C/1000m (.2°C/100m).

1. In the following problem, the temperature of a parcel of air at the ground is 30°C, which is 6°C warmer than the surrounding air (24°C). The dew point is 6°C. The warm parcel of air begins to rise and cool. Enter temperature of the rising parcel of air and the surrounding air, and dew point for each elevation on the table.

   The parcel of air will rise only if it is warmer than the surrounding air. At the elevation where the dew point equals the temperature of the rising air, the relative humidity is 100 percent.

Elevation (m)	Temperature (°C) ELR	Temperature (°C) DAR	Dew Point (°C)
4000-	−4	−10	−2
3500-	−0.5	−5	−1
3000-	3	0	0
2500-	6.5	5	1
2000-	10	10	2
1500-	13.5	15	3
1000-	17	20	4
500-	20.5	25	5
surface-	24	30	6
	↑ surrounding air	↑ rising air	

In this problem, the relative humidity (and therefore dew point) at the surface was low. If your calculations are correct, the column of air stopped rising before reaching the dew point, and no condensation or precipitation occurred.

   a. How high (or to what elevation) did the air parcel rise?

   _____ m

   b. How much higher did the parcel of air need to rise to reach the dew point?

   _____ m

2. In the following problem, the temperature at the surface is 30°C, which is the same as in the previous problem. The only difference is that humidity (and dew point) at the surface is higher. Higher humidity increases the probability that the rising air will reach the condensation level.

   Calculate the temperature of a rising parcel of air. As in the previous problem, the air will only rise if it is warmer than the surrounding air.

   Calculate and enter temperature and dew point on the following table. Remember to use the saturated adiabatic rate if the relative humidity is 100 percent.

	Temperature (°C)		
Elevation (m)	ELR	DAR	Dew Point (°C)
4000-	_____	_____	_____
3500-	_____	_____	_____
3000-	_____	_____	_____
2500-	_____	_____	_____
2000-	_____	_____	_____
1500-	_____	_____	_____
1000-	_____	_____	_____
500-	_____	_____	_____
surface-	24	30	18
	↑	↑	
	surrounding air	rising air	

a. At what elevation was dew point reached?

_____ m

b. What is the difference in temperature between the rising column of air and the surrounding air at:

2000m? _____

3000m? _____

4000m? _____

c. Note that after reaching the condensation level, the temperature of the rising air becomes progressively warmer than the surrounding air. Why is the temperature difference greater as the air rises (what is the heat source)?

d. Based on your findings in the last two problems, why is convectional precipitation much more common in humid as opposed to desert regions?

3. In this problem you will measure temperature and humidity immediately outside your building to determine the level of condensation, or the cloud base.
Use a sling psychrometer over a vegetated surface that is in full sunlight to determine:

dry bulb temperature _____ °C

wet bulb temperature _____ °C

wet bulb depression _____ °C

Use Table 8.2 to determine relative humidity: _____ %

Use Figure 8.1 to determine the dew point: _____ °C

Enter the temperature and dew point at the surface on the following table. Calculate and enter these values for the elevations given.

Elevation (m)	Temperature (°C)	Dew Point (°C)
5000-	_____	_____
4500-	_____	_____
4000-	_____	_____
3500-	_____	_____
3000-	_____	_____
2500-	_____	_____
2000-	_____	_____
1500-	_____	_____
1000-	_____	_____
500-	_____	_____
surface-	_____	_____

a. What is the elevation of the cloud base?

_____ m

b. What is the dew point at the level of condensation?

_____ °C

On Figure 9.3 plot a line showing the temperature decrease with increasing elevation. Remember to use the DAR if the relative humidity is less than 100 percent, and the SAR once the air has become saturated. Also, draw a line showing the decrease in dew point with increasing elevation (2°C/1000m). The elevation where the temperature and dew point lines meet indicates the elevation of the cloud base (relative humidity is 100 percent). Label all lines and mark the level of condensation.

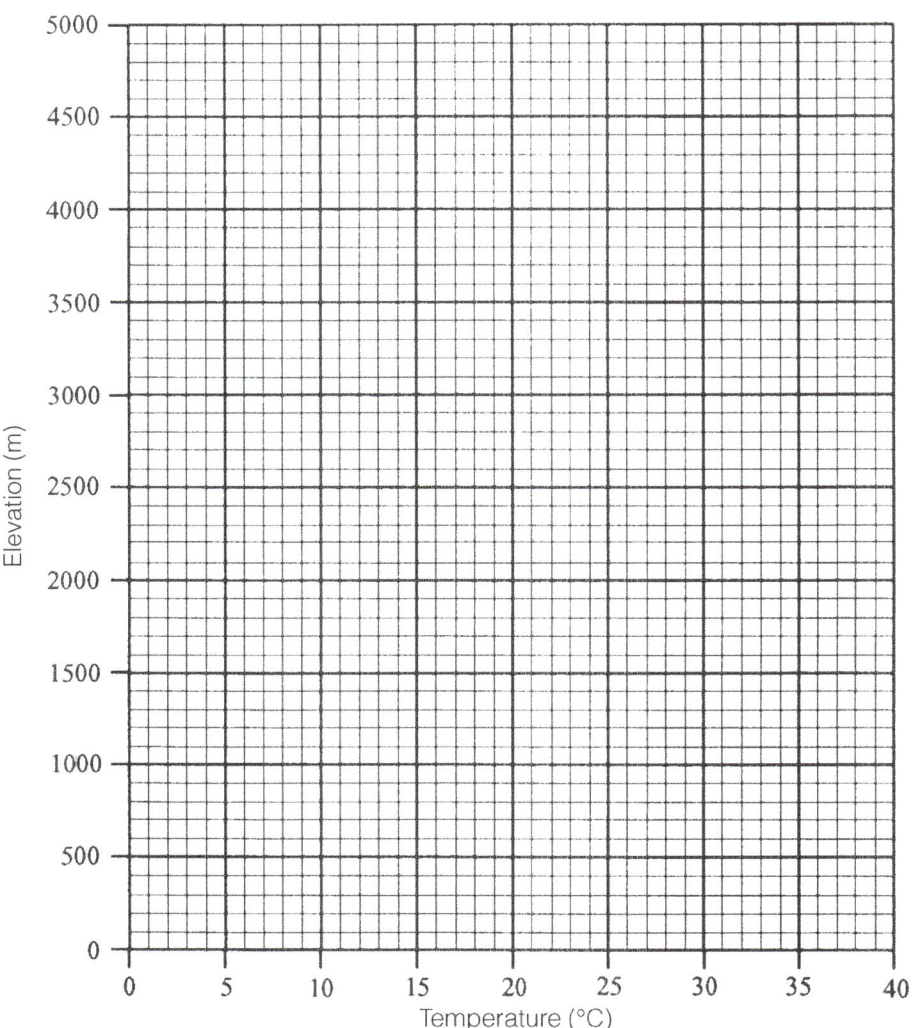

FIGURE 9.3 Temperature decrease with increasing elevation.

Name_____ Section_____

Air Masses, Cyclonic Storms, and Fronts

LAB EXERCISE 10

Air masses and pressure systems cover extensive areas. Viewing atmospheric conditions over a large geographic region is necessary to understand weather at any single position. The objective of this exercise is for you to (1) become familiar with surface weather maps, and (2) to analyze weather systems and make short-term weather forecasts based on these maps.

The National Weather Service collects surface weather data from dozens of weather recording stations around the country. Every day new data are plotted on maps to show large scale atmospheric phenomena such as air masses, cyclonic storms, fronts, winds, cloud cover, and precipitation. The location of each recording station is indicated by a circle on the map. The weather station model (Figure 10.1) refers to symbols around this circle indicating weather conditions at the time of data collection. Most weather stations do not report all of the information shown

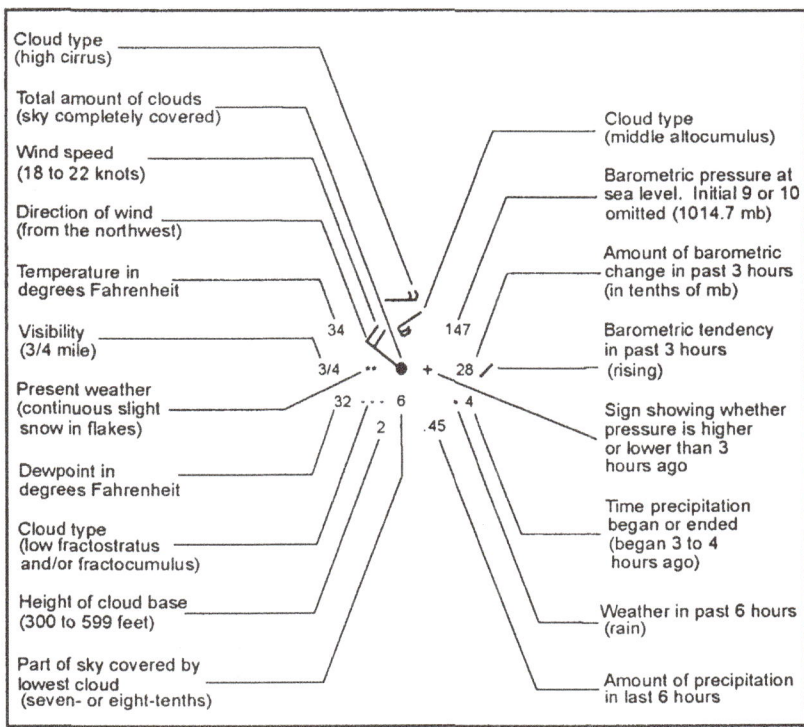

FIGURE 10.1 Weather station model. *(Source: National Weather Service.)*

95

on this sample. But air temperature, dew point temperature, and barometric pressure are almost always included. Note that pressure is indicated by the last three digits of millibar readings. A 9 or 10 must be placed in front of the 3 digits. For example, a pressure reading of 1024.6 mb would be shown as 246.

The amount of cloud cover is represented by the type of shading inside the circle (Figure 10.2). Wind direction is shown by the wind arrow, or line that indicates the direction of wind toward that station. The tails or feathers on the end of the wind arrow indicate wind speed. The weather map symbols used on these maps are shown below in Figure 10.2.

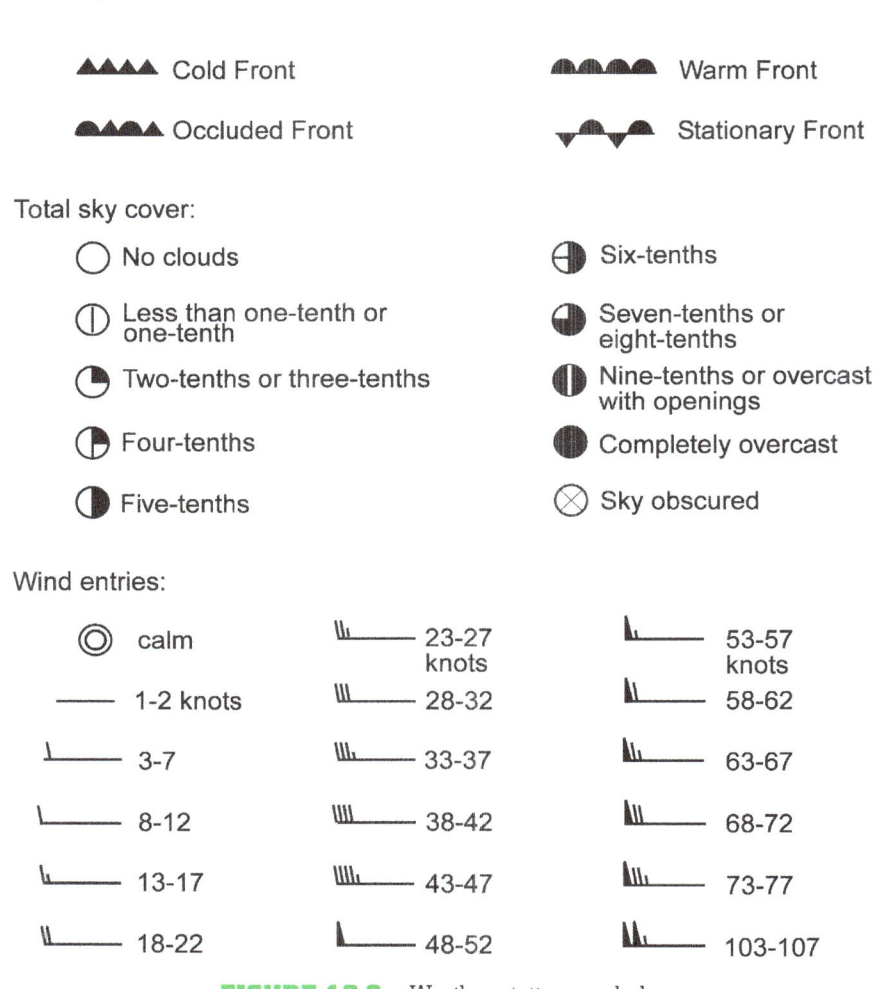

FIGURE 10.2 Weather station symbols.

1. Figure 10.3 is an example of a weather map showing atmospheric conditions on February 11, 1988. An area of low pressure is centered over Kentucky (Lat. 37°N, Long. 86°W). Counterclockwise circulation around the cyclonic storm has moved cold air unusually far to the south. Note the air mass boundaries indicating the edge of the cold air mass. These are shown by a cold front extending from Kentucky to northern Mexico and a stationary front extending from Canada south to New Mexico.

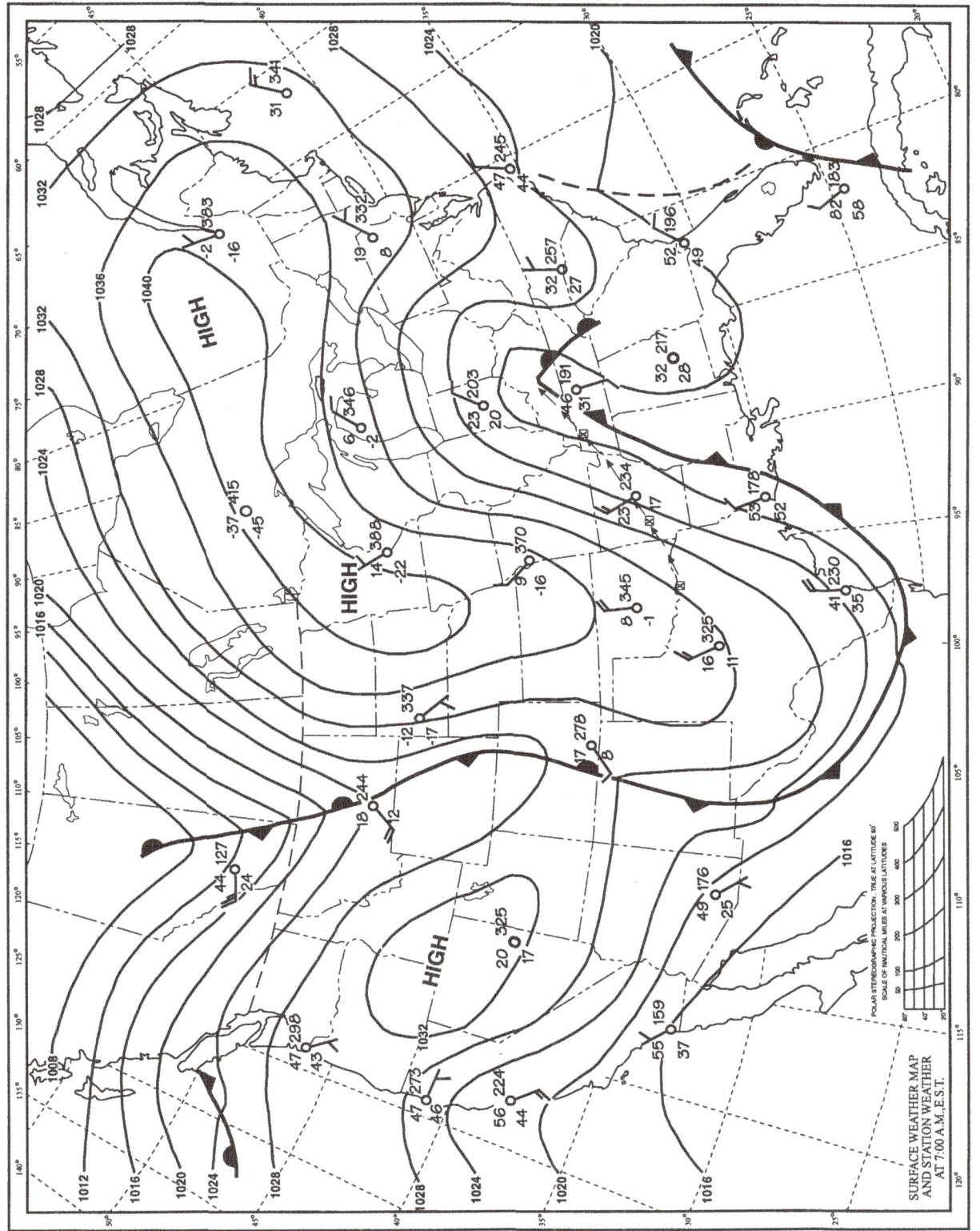

FIGURE 10.3 Surface weather map for February 11, 1988.

97

a. What are the highest and lowest barometric pressure readings (station data) on the map and where are they located?

high pressure: _____

location: _____

low pressure: _____

location: _____

b. What is the coldest temperature on the map (station data) and where is it located?

low temperature: _____

location: _____

c. What is the wind direction and speed in central Oklahoma?

d. Compare the temperature at San Francisco to stations in the following states that are at about the same latitude (35°N):

Nevada

Oklahoma

North Carolina

e. Why is winter temperature at the California coast warmer than most other areas in the country?

f. Referring to the weather station closest to you, give the following information:

Temperature: _____

Wind direction: _____

Wind speed: _____

Barometric pressure: _____

2. Two weather maps show Hurricane Andrew in the Gulf of Mexico on August 24, 1992 and two days later, August 26, 1992 (Figures 10.4 and 10.5). Andrew crossed southern Florida causing extensive damage, and less than two days later hit the coast of Louisiana. Note that Andrew moved generally east to west following the trade winds. In the central U.S. a cold front is moving in the opposite direction because it is at a high latitude and therefore in the zone of the westerlies.

   a. What is the latitude and longitude of Andrew on:

   August 24? _____

   August 26? _____

   b. What is the average forward speed that the center of the hurricane is moving between August 24 and 26?

   _____

   c. In which direction do winds circulate around the hurricane?

   _____

   d. What is the barometric pressure in the center of Andrew on August 24?

   _____

   e. On August 26, the highest barometric pressure on the map (station data) is

   _____ mb

   location: _____

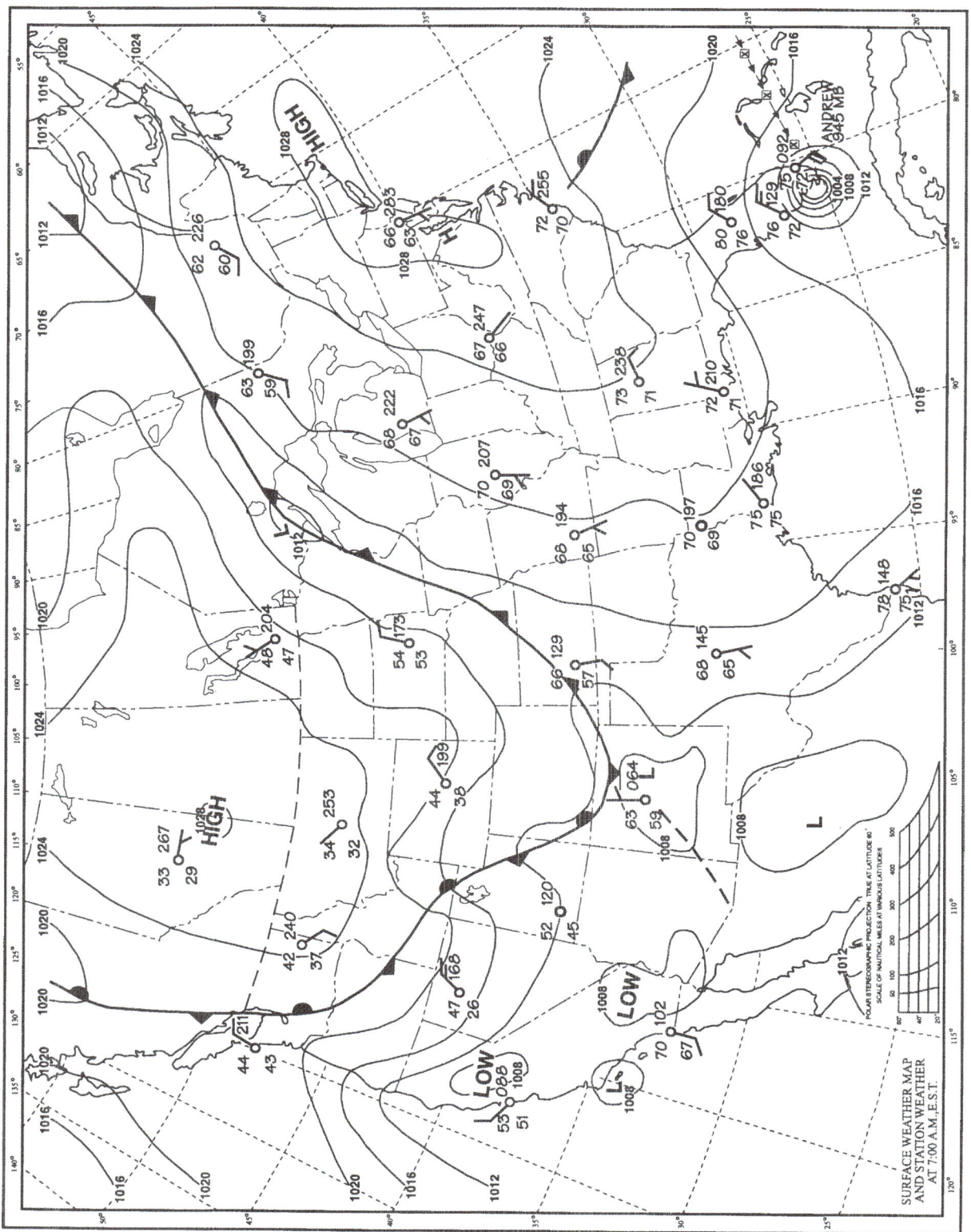

FIGURE 10.4 Surface weather map for August 24, 1992.

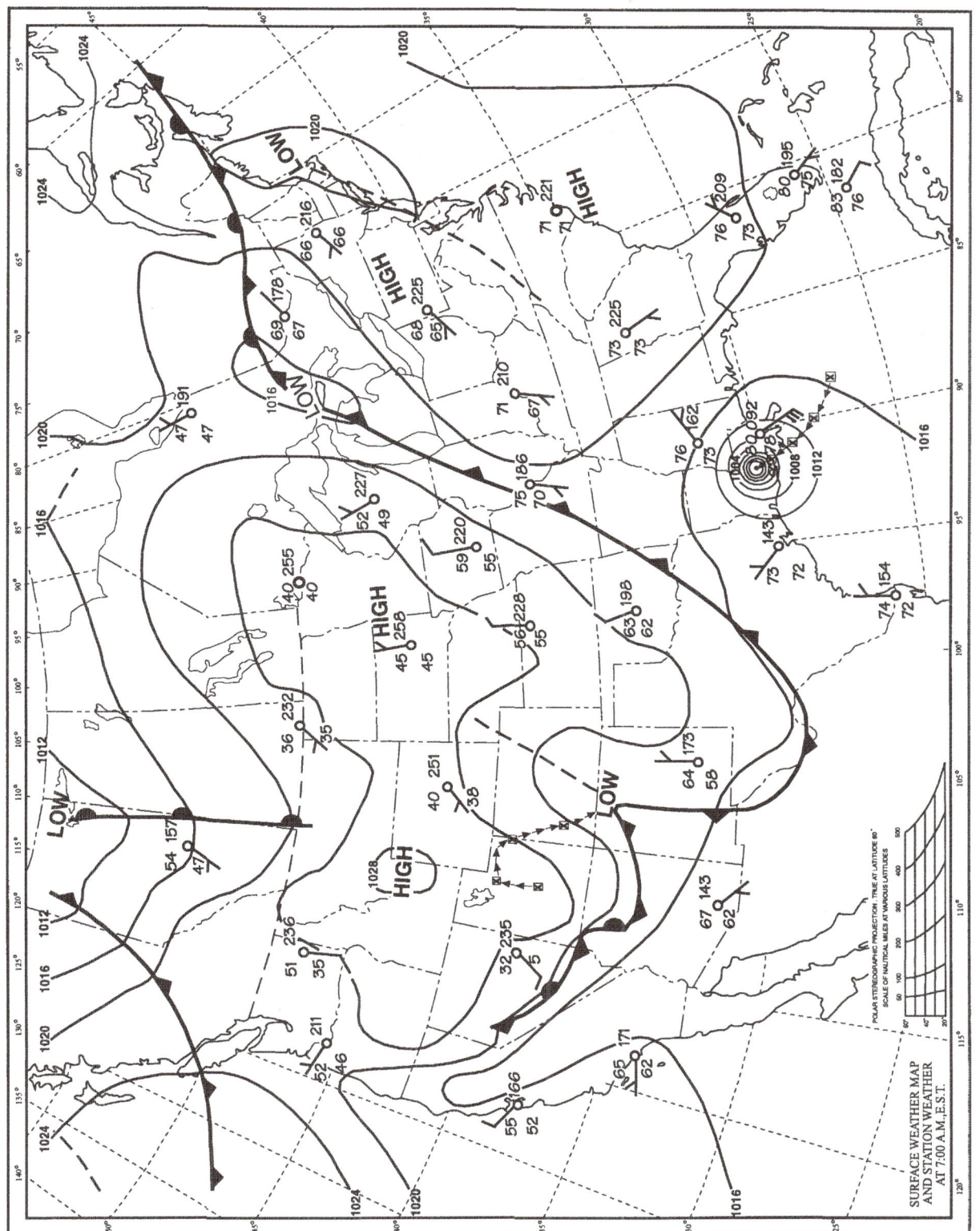

FIGURE 10.5 Surface weather map for August 26, 1992.

f. Andrew moved inland and then dissipated. Why do hurricanes lose power over land?

g. Describe the change of the position of the cold front in the central U.S. between August 24 and 26.

h. Generally, what are the differences in air temperature on either side of the cold front?

3. By looking at daily weather maps over a period of several days, you can easily track the movement of atmospheric phenomena, such as cyclonic storms, high pressure, and fronts. A series of four weather maps (Figures 10.6–10.9) covering the period of October 16–19, 1990 shows changing meteorological conditions through time. The map of October 16 shows low pressure in the western U.S. and high pressure located in the eastern U.S. The cyclonic storm (low pressure) strengthens and begins to move to the east. The movement of the storm is indicated by a chain of arrows. Associated with the storm are well defined cold and warm fronts.

   a. What is the average speed that the center of the cyclonic storm was moving between October 16–19?

   _____

   b. Describe the change in position of the cold front during this 4-day period.

   c. What was the general wind direction and speed in the region of the Ohio River valley on October 16 and 18?

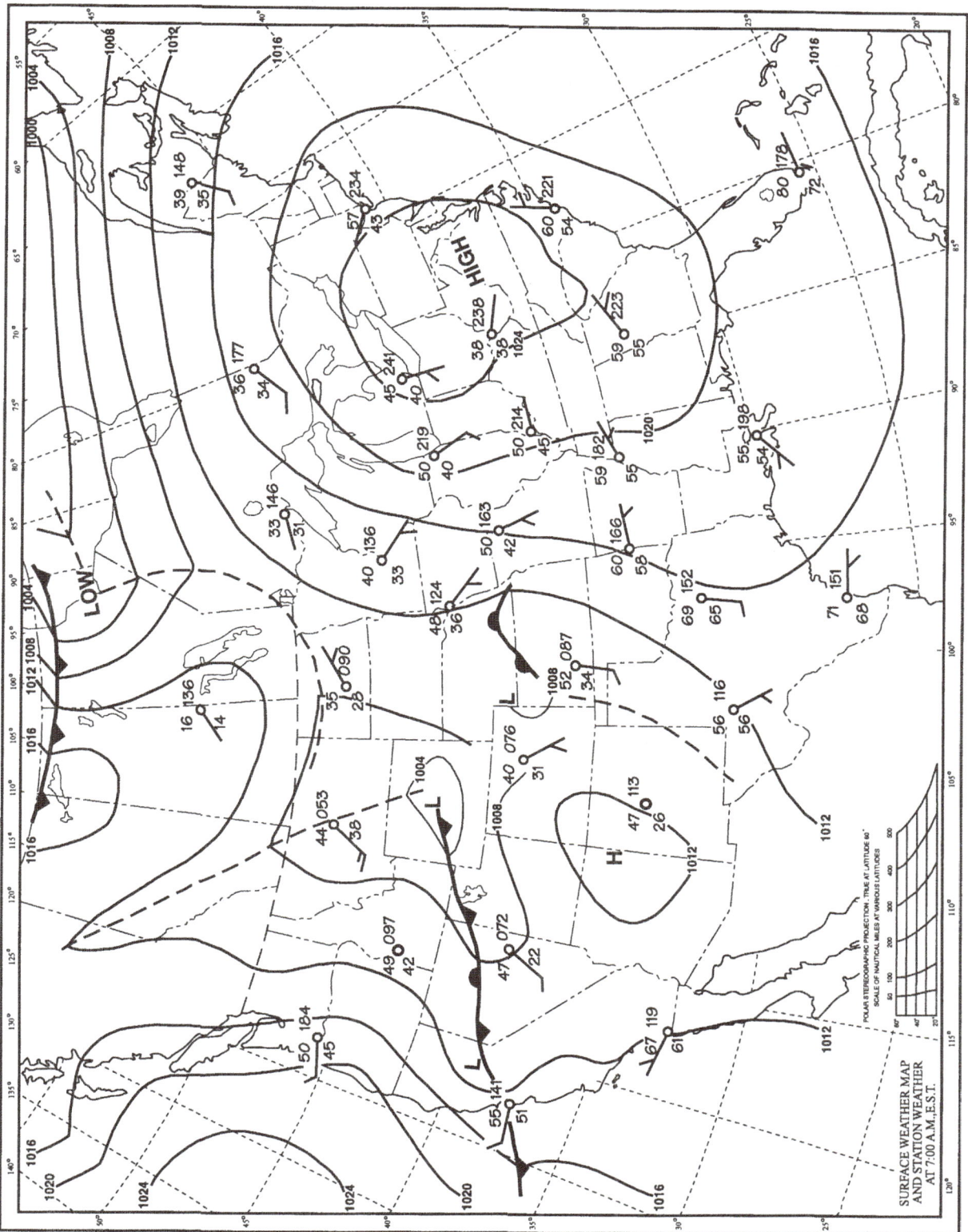

FIGURE 10.6 Surface weather map for October 16, 1990.

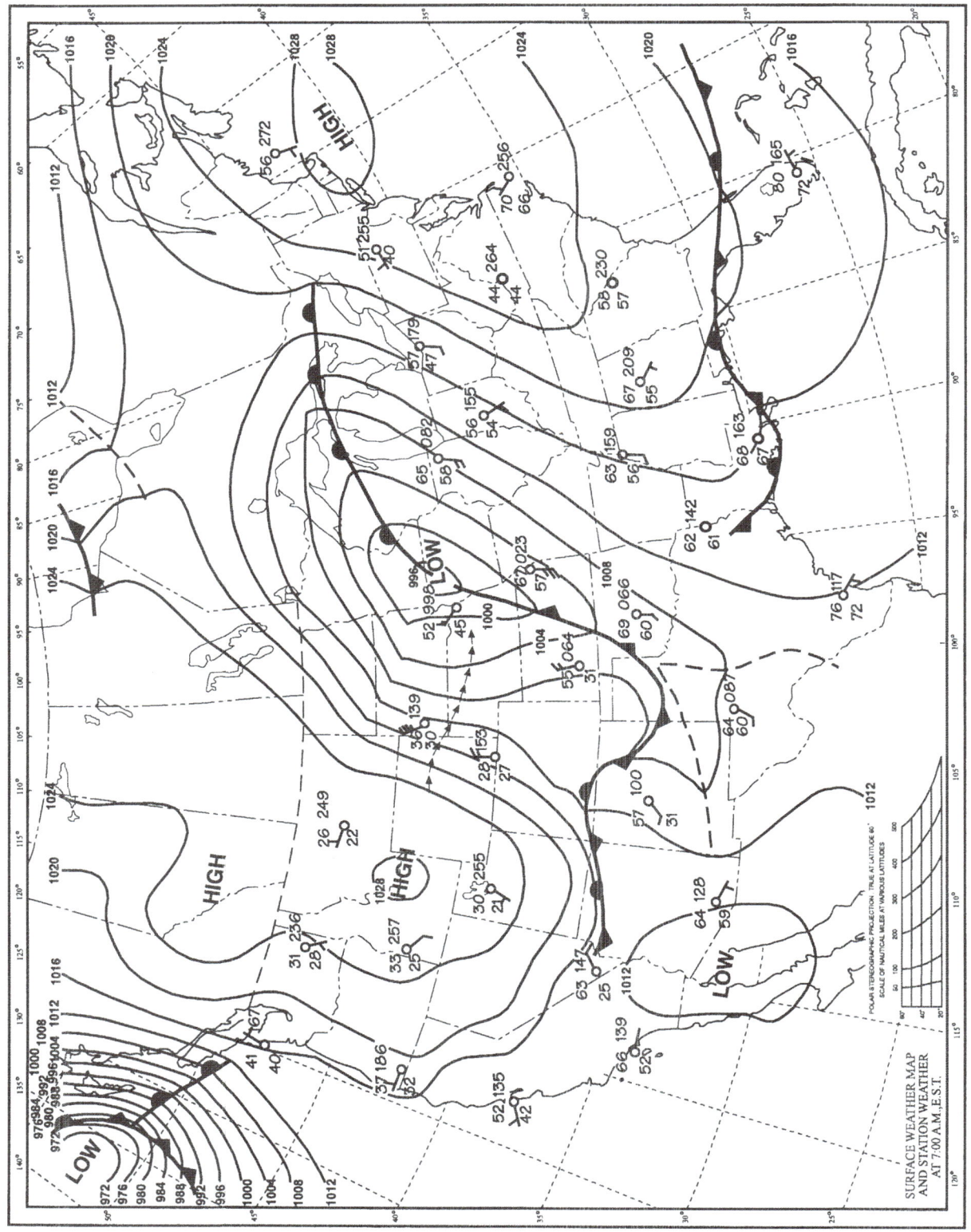

FIGURE 10.7 Surface weather map for October 17, 1990.

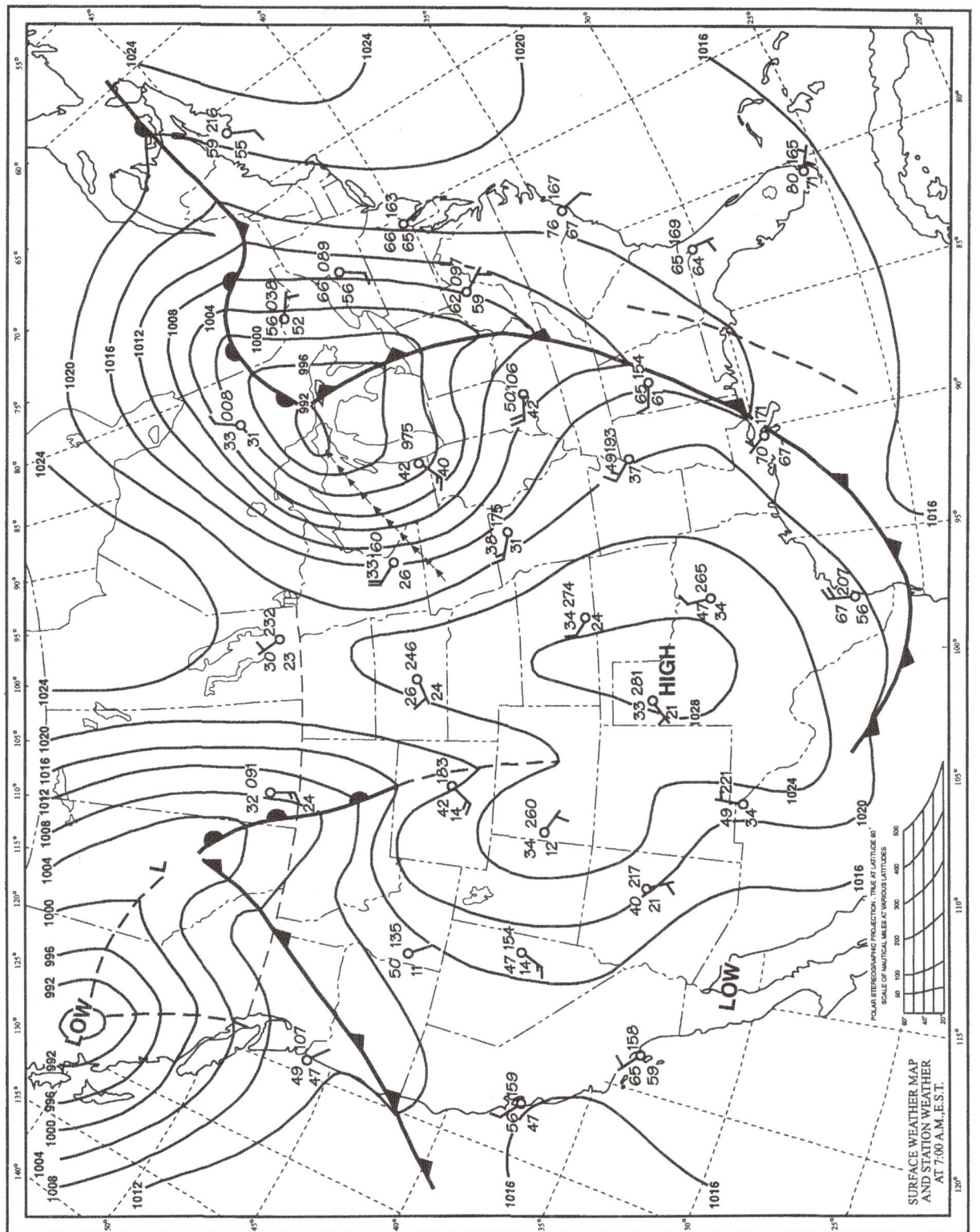

FIGURE 10.8 Surface weather map for October 18, 1990.

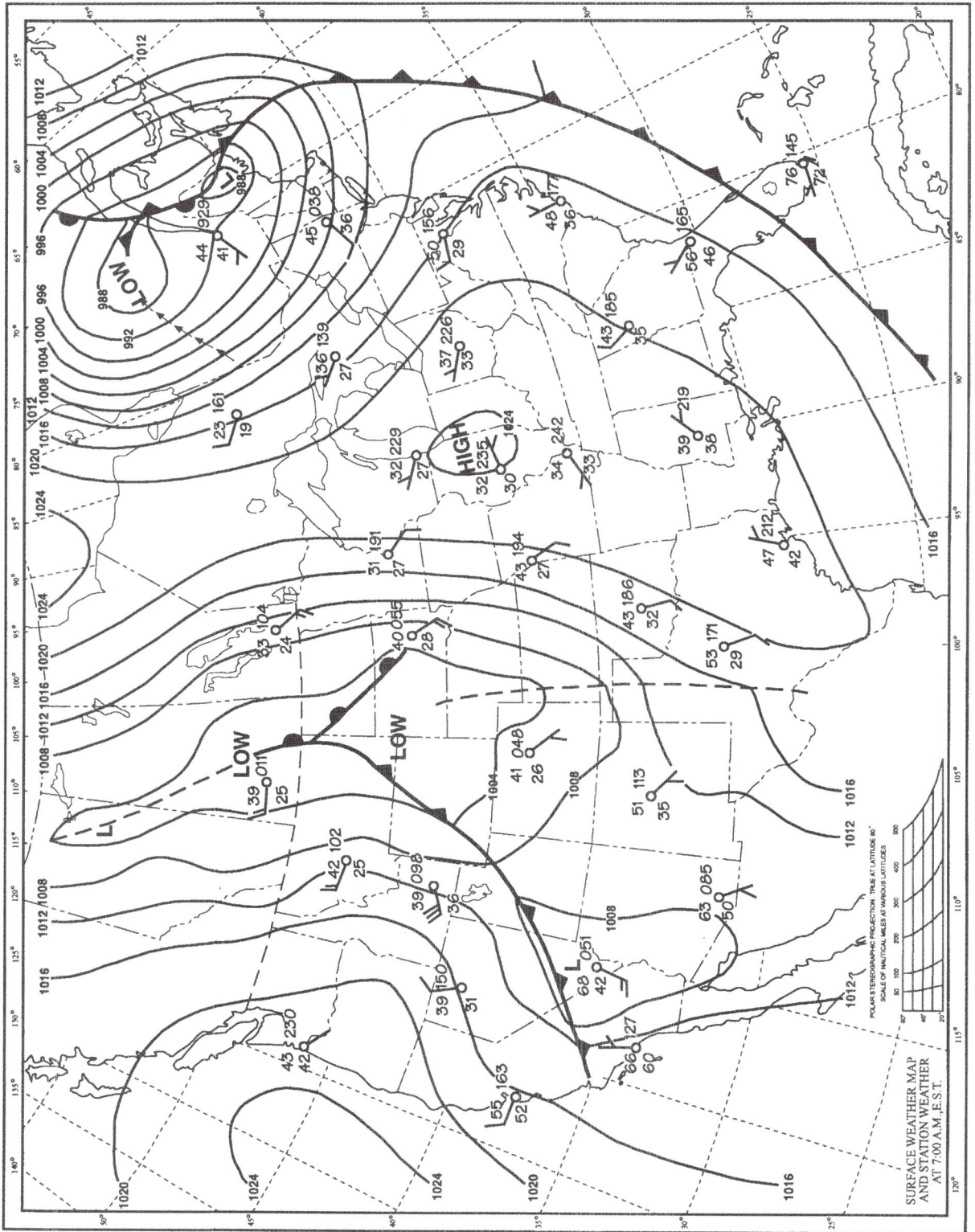

FIGURE 10.9 Surface weather map for October 19, 1990.

d. The center of the cyclonic storm passed over Lake Michigan. Use the isobars to determine the changes in barometric pressure during this four day period.

October 16: _____ mb

October 17: _____ mb

October 18: _____ mb

October 19: _____ mb

e. What is the relationship between areas of precipitation and the position of high pressure, low pressure, and fronts?

Referring to the weather station closest to you:

(1) Give the following information for October 16 based on Hattiesburg, MS (Figure 10.6):

Temperature: _____

Wind direction: _____

Wind speed: _____

Barometric pressure: _____

(2) Refer to the maps showing conditions during the next three days, October 17–19 (Figures 10.7–10.9). Describe how the conditions listed above changed during this period. Also describe the relationships between these changes and the movement of the cyclonic storm and high pressure, and the passage of fronts.

Name _____ Section _____

Tropical Cyclones and Hurricanes

LAB EXERCISE 11

Tropical cyclones are one of the most destructive natural forces on the planet, capable of near catastrophic damage. The Gulf and Atlantic coasts of North America, and the Caribbean coast of Central America, are no strangers to these storms, which commonly form in the warm waters of the tropical North Atlantic Ocean and migrate towards these areas in an easterly fashion. When these storms organize and obtain core wind speeds of 74 miles per hour, which is significant enough to form an eye, they are typically given the moniker—Hurricane.

In the previous chapter, we briefly discussed the pressure and circulation surrounding Hurricane Andrew, which made its final landfall near Franklin, Louisiana on August 26, 1992. In this chapter we will take a more in-depth look at the geography of these events by focusing on the record breaking 2005 Atlantic hurricane season, which boasted 15 hurricanes over a period of six months.

Hurricane Development

All hurricanes begin life as an undeveloped thunderstorm. If these thunderstorms are located in a region that supports further development, over time they may be able to become better organized and strengthen into a hurricane. Hurricanes typically form over warm ocean water that is 21°C (80°F) or greater, and where evaporation and lifting are great. Evaporation is important for the latent heat release from condensation that is occurring at the top of the storm. The hurricane uses the energy released from this process as fuel to sustain itself or for further strengthening. In addition to these criteria, strong upper-level winds cannot be present. These winds have the ability to shear off the top of a hurricane and cause significant weakening.

The majority of Atlantic hurricanes form out of **easterly waves**. Easterly waves are series of wave-like oscillations (alternating zones of high and low pressure) that are embedded in the trade winds. These powerful thunderstorms can develop in the low pressure sections of these waves and strengthen very rapidly into hurricanes. Once developed, they travel to the west with the trade winds. Hurricanes that develop from easterly waves and do not receive support or enhancement from anywhere else are known as **Tropical-Only (TO) Hurricanes**.

The other major type of Atlantic hurricane is known as the **Baroclinically-Enhanced (BE) Hurricane**. A baroclinic zone is a region where two or more distinct airmasses are converging, which results in an unstable atmosphere and frontal lifting. These baroclinic zones usually de-

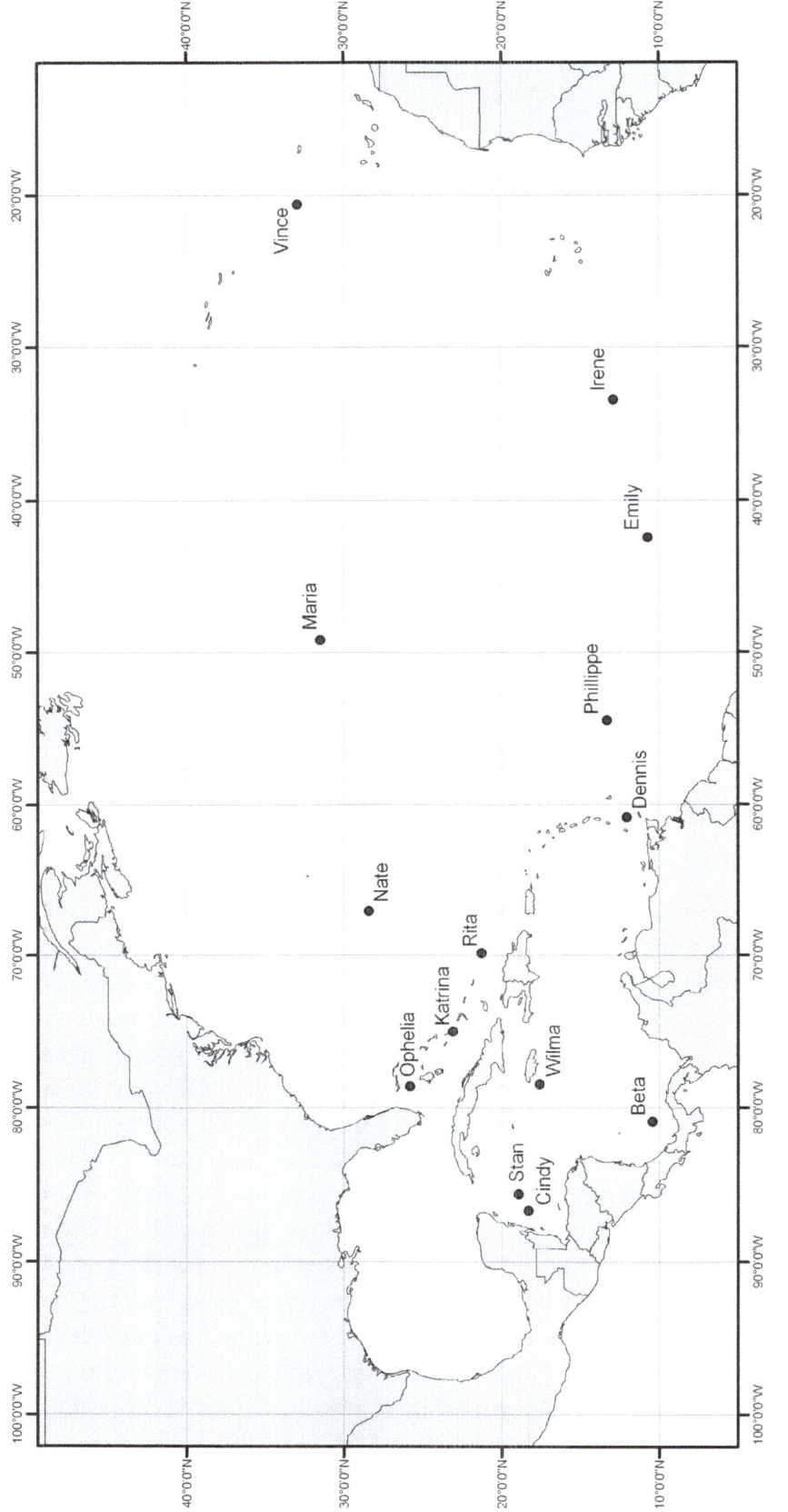

FIGURE 11.1 Map of the Atlantic basin showing the point of genesis for the 15 hurricanes which formed during the 2005 hurricane season.

veloped outside the tropics in the mid-latitudes. If a thunderstorm develops inside a baroclinic zone, the unstable atmosphere can help them strengthen into a hurricane very rapidly. BE hurricanes usually form further to the north (mid-latitudes), away from easterly waves and the tropics. Also, hurricanes that start as tropical-only but are then influenced by a baroclinic zone are classified as BE hurricanes.

Use the map in Figure 11.1 to answer the following questions:

1. Did the majority of the 2005 hurricanes form inside or outside the tropics?

    _____

2. List the names of 4 hurricanes, which occurred during the 2005 season, that are most likely to be TO hurricanes based on their location of genesis. Note: for a more accurate assessment, go to the National Hurricane Center's website at http://www.nhc.noaa.gov and look up each of the hurricane tracks in the 2005 season archive.

    _____   _____

    _____   _____

3. List 4 hurricanes that are most likely to be BE hurricanes based on their location of genesis.

    _____   _____

    _____   _____

4. In the 2005 data, is there a noticeable pattern to the locations of hurricane genesis? For example, do early season storms tend to form in a certain location, whereas mid-season or late season storms develop elsewhere.

Hurricane Movement

Figure 11.2 shows the tracks of every hurricane that occurred in the Atlantic basin during 2004. Though there is quite a bit of variability when comparing individual hurricanes, the average hurricane track is easy to identify. Most hurricanes develop within the tropics and then migrate to the west-northwest within the trade winds. As these storms approach the mid-latitudes, they are bent back to the northeast as they begin to travel with the mid-latitude westerly flow.

Aside from the Earth's general wind belts, the Bermuda High greatly affects the tracks of Atlantic Hurricanes. The Bermuda High is a subtropical, semi-permanent zone of high pressure that is typically located over the western North Atlantic basin between 25 and 35 degrees north latitude. Because zones of high pressure spin clockwise in the Northern Hemisphere, this circulation helps to curve the path of hurricane travel in a clockwise fashion (Figure 11.3).

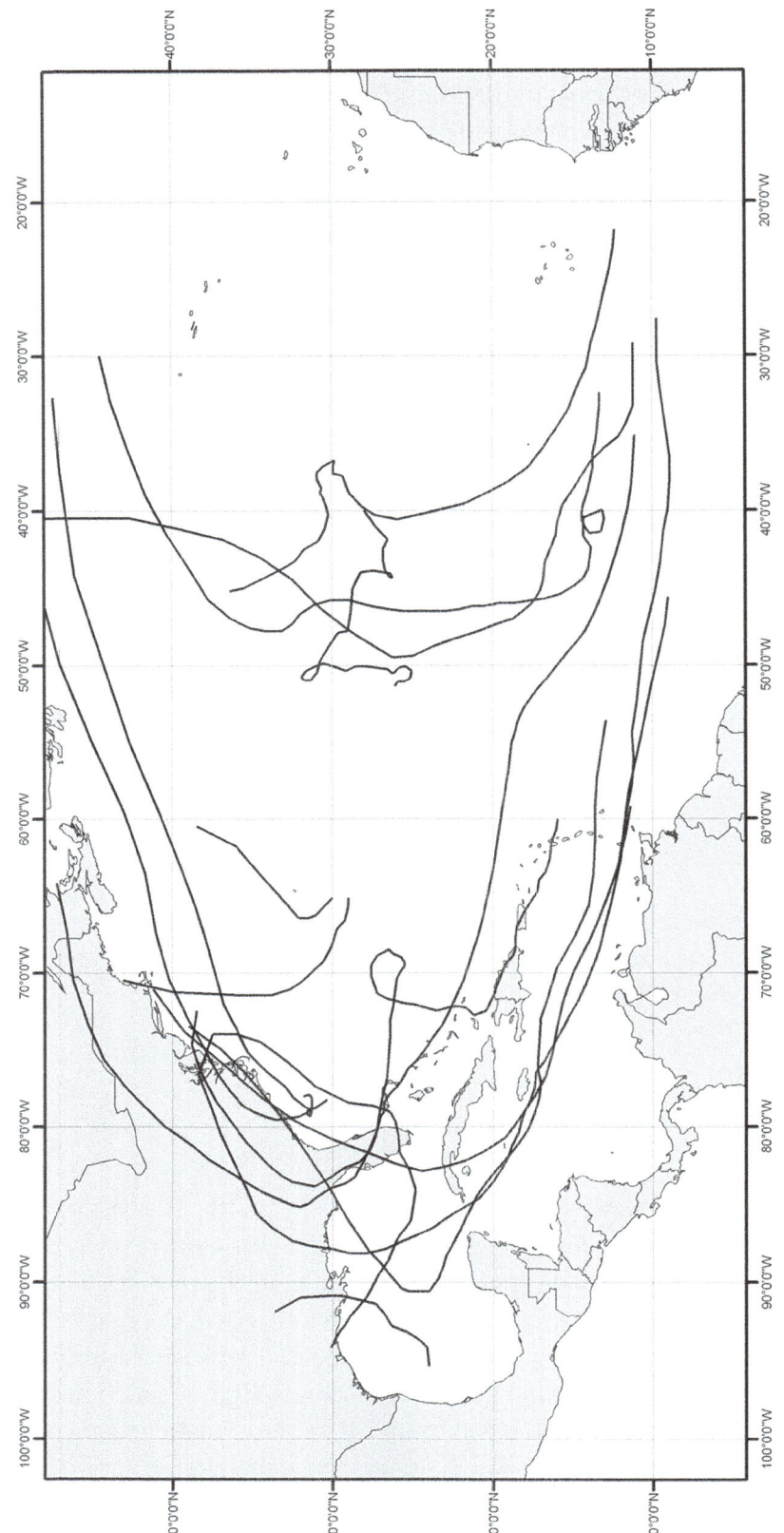

FIGURE 11.2 Hurricane tracks during 2004 in the Atlantic Basin.

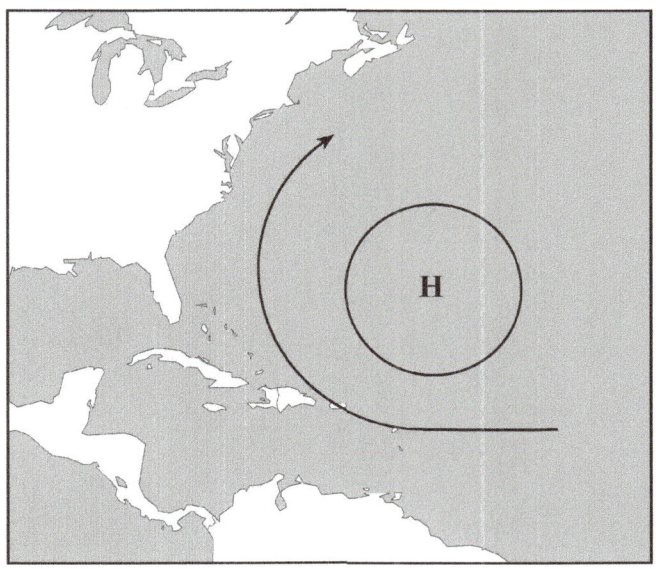

FIGURE 11.3 Diagram of the Bermuda High and its influence on a hypothetical hurricane track.

It is important to note that the Bermuda High is not a static feature, and shifts location on a variety of different time-scales. The position of the Bermuda High can have an enormous impact on the track of a hurricane and the potential area of landfall.

1. One the three diagrams below, draw a circle around the Bermuda High, indicate its direction of rotation, and draw in the track of a hypothetical hurricane as it moves around the Bermuda High (as in Figure 11.3).

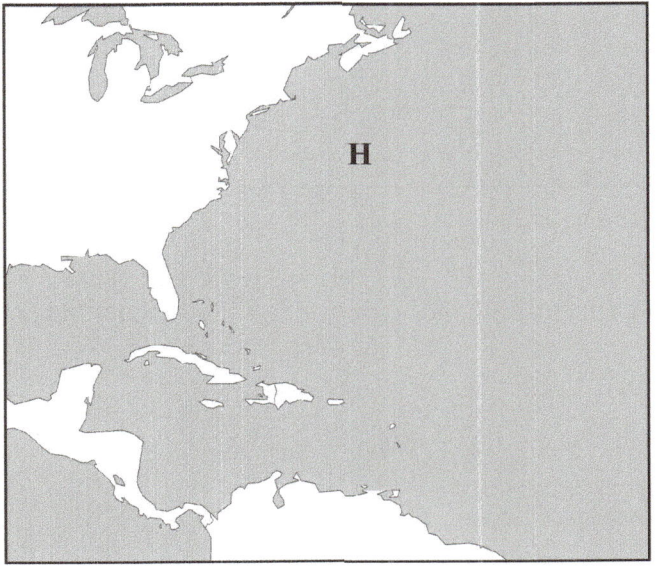

1a.

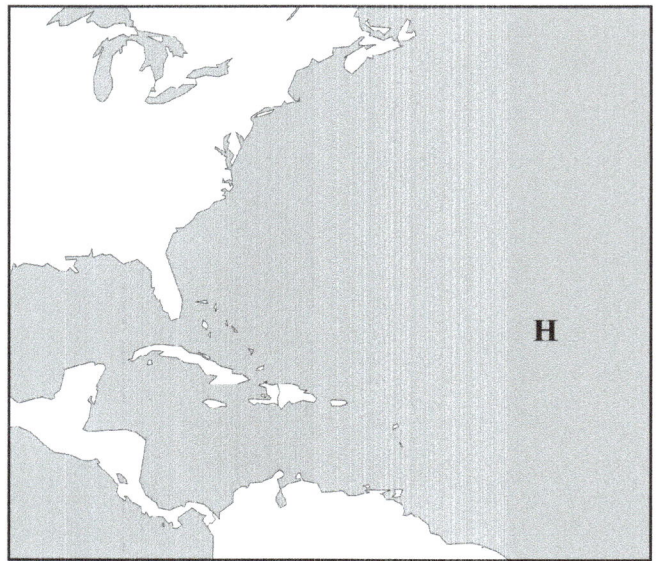

1b.

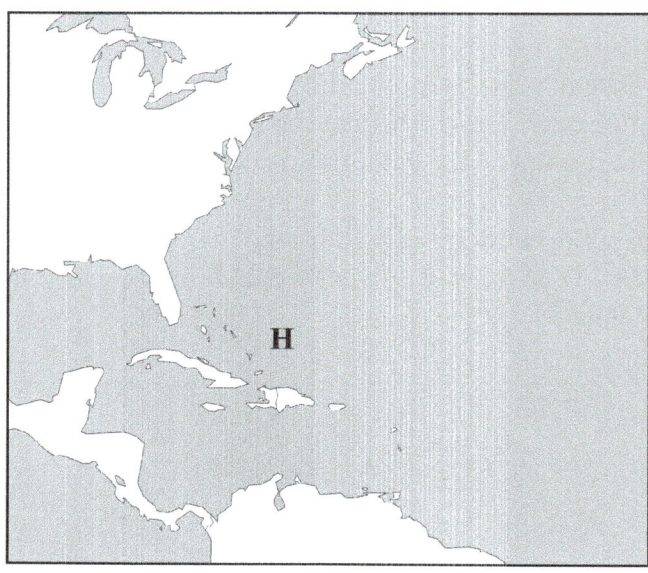

1c.

2. How does changing the position of the Bermuda High, impact the potential point of landfall?

3. At what position (1a, 1b, or 1c) are the residents of the Gulf of Mexico Coast in greatest danger? _____

4. At what position (1a, 1b, or 1c) are the residents of the Atlantic seaboard in greatest danger?

   _____

5. If we knew that the Bermuda High was going to remain in a location similar to that in diagram 1c, for the next 100 years, what would that do to coastal property insurance rates in the following locations; (1) New Orleans, Louisiana, (2) Virginia Beach, Virginia, and (3) Cozumel, Mexico.

Hurricane Tracking

Hurricane Katrina made landfall near the end of August during the 2005 hurricane season, and was the costliest natural disaster in the history of the United States. Table 11.1 contains the coordinates of the storm as it tracked across the Atlantic basin between the 23rd and the 30th of August, 2005.

1. Using this data plot the course of the storm (all 30 points) over this 7-day period on the hurricane tracking chart provided on page 117. Also using the Saffir-Simpson scale provided for you (Table 11.2), fill in the category of the storm for each entry in Table 11.1 and label each point on the tracking chart with the appropriate category.

TABLE 11.1 TRACKING COORDINATES FOR HURRICANE KATRINA, AUGUST 23–30, 2005

Time	Date	Latitude	Longitude	Wind Speed (mph)	Pressure (mb)	Category
1 pm	8/23/05	23.1° N	75.1° W	35	1008	
7 pm	8/23/05	23.4° N	76.2° W	35	1007	
1 am	8/24/05	23.8° N	76.2° W	35	1007	
7 am	8/24/05	24.5° N	76.5° W	40	1006	
1 pm	8/24/05	25.4° N	76.9° W	45	1003	
7 pm	8/24/05	26.0° N	77.7° W	50	1000	
1 am	8/25/05	26.1° N	78.4° W	60	997	
7 am	8/25/05	26.2° N	79.0° W	65	994	
1 pm	8/25/05	26.2° N	79.6° W	70	988	
7 pm	8/25/05	25.9° N	80.3° W	80	983	
1 am	8/26/05	25.4° N	81.3° W	75	987	
7 am	8/26/05	25.1° N	82.0° W	85	979	
1 pm	8/26/05	24.9° N	82.6° W	100	968	
7 pm	8/26/05	24.6° N	83.3° W	105	959	
1 am	8/27/05	24.4° N	84.0° W	110	950	
7 am	8/27/05	24.4° N	84.7° W	115	942	
1 pm	8/27/05	24.5° N	85.3° W	115	948	
7 pm	8/27/05	24.8° N	85.9° W	115	941	
1 am	8/28/05	25.2° N	86.7° W	145	930	
7 am	8/28/05	25.7° N	87.7° W	165	909	
1 pm	8/28/05	26.3° N	88.6° W	175	902	
7 pm	8/28/05	27.2° N	89.2° W	160	905	
1 am	8/29/05	28.2° N	89.6° W	145	913	
7 am	8/29/05	29.5° N	89.6° W	125	923	
1 pm	8/29/05	31.1° N	89.6° W	90	948	
7 pm	8/29/05	32.6° N	89.1° W	60	961	
1 am	8/30/05	34.1° N	88.6° W	45	978	
7 am	8/30/05	35.6° N	88.0° W	35	985	
1 pm	8/30/05	37.0° N	87.0° W	35	990	
7 pm	8/30/05	38.6° N	85.3° W	35	994	

Atlantic Basin Hurricane Tracking Chart

National Hurricane Center, Miami, Florida

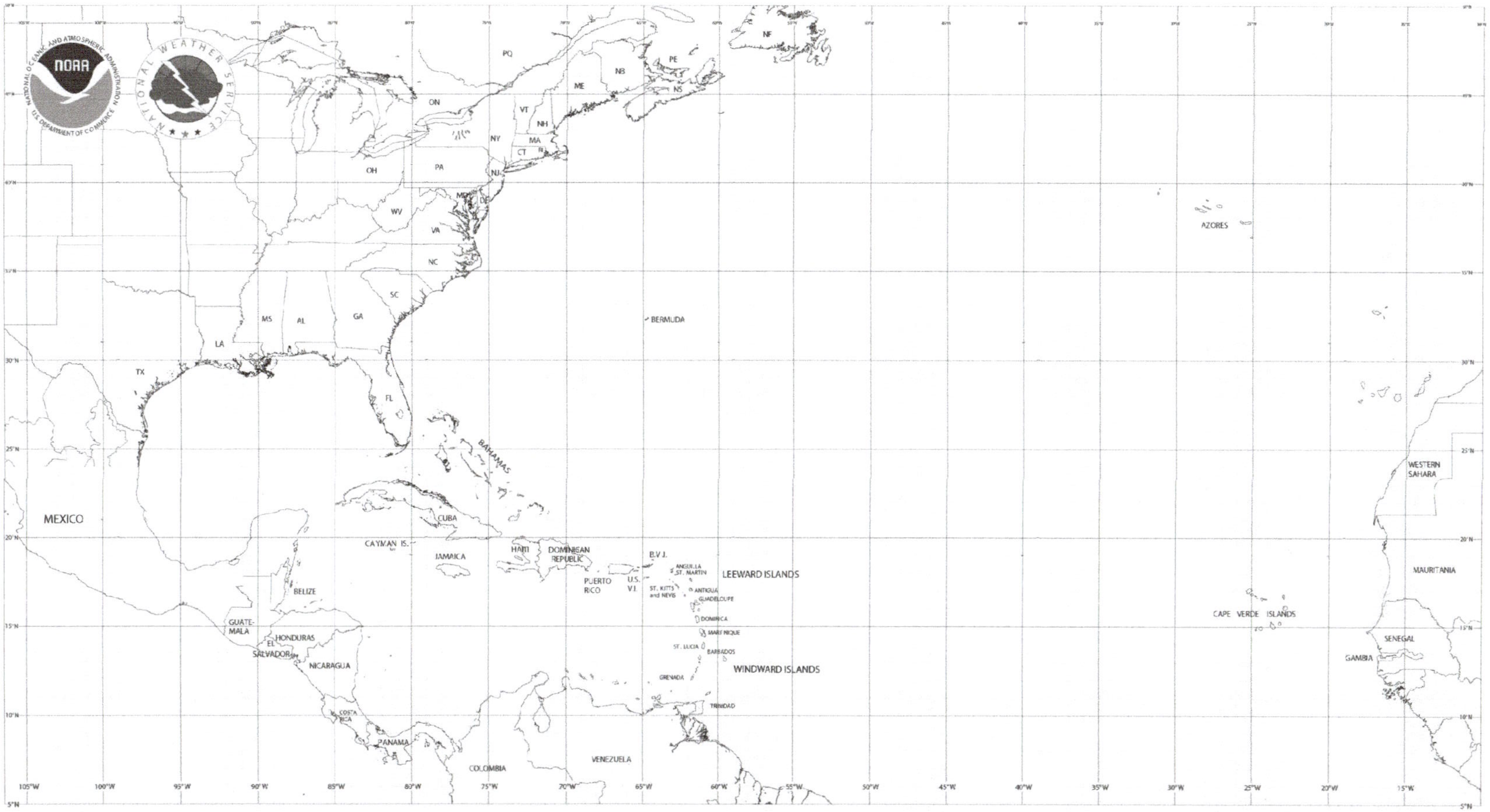

FIGURE 11.4 Hurricane tracking map.

TABLE 11.2. THE SAFFIR SIMPSON SCALE

Storm Type	Maximum Sustained Winds	
	Knots	Mph
Tropical Depression	Less than 34	Less than 39
Tropical Storm	34–63	39–73
Category 1 Hurricane	64–82	74–95
Category 2 Hurricane	83–95	96–110
Category 3 Hurricane	96–113	111–130
Category 4 Hurricane	114–135	131–155
Category 5 Hurricane	136+	156+

2. How many landfalls did Hurricane Katrina have? _____

3. Where did Katrina make landfall, and what category was the hurricane at each landfall?

4. Between which two time periods did Katrina experience the greatest change in pressure?

   _____

5. Between which two time periods did Katrina experience the greatest change in wind speed?

   _____

6. What was happening during these time periods (in questions 4 and 5) and why would it affect the hurricane so rapidly?

Landfall and Storm Surge

As a hurricane moves across the ocean, its winds and general forward motion push a large amount of surface water out in front of the storm. The sea level rise associated with this "piling up" of water ahead of the storm is referred to as the storm surge. The height of the storm surge is highly correlated with the strength of the storm; in general the higher the sustained wind speeds, the higher the storm surge. However, the storm surge is not uniform across the entire hurricane. The right-front quadrant of the hurricane, where the counter-clockwise internal rotation of the hurricane is in synch with the forward motion, is typically the highest. In this quadrant the surge is being directed straight at the land, as opposed to the left-front quadrant, where the rotation of the storm is directing the surge away from the land. Although both front quadrants will experience some amount of storm surge as the hurricane makes landfall, the left-front will be greatly reduced compared to the right-front (Figure 11.5). In addition to this, the shape of the land can also have an affect on storm surge heights. Concave-shaped coasts (like bays) tend to funnel the storm surge waters making the realized surge height higher than normal. Convex-shaped coastlines have the opposite effect, dispersing storm surge waters away from the point of impact.

FIGURE 11.5   The motion of a hurricane and the resulting relative storm surge levels.

As a hurricane begins to make landfall, it has less surface area over the ocean and therefore less access its fuel source; water vapor. This causes the hurricane to lose strength just prior to landfall, and then rapidly deteriorate once landfall has been made. Likewise, the storm surge that was pushed ashore by the hurricane tends to recede very quickly once landfall has been made. Though the process of landfall is relatively quick, hurricanes are capable of near catastrophic damage, as was the case of Hurricane Katrina.

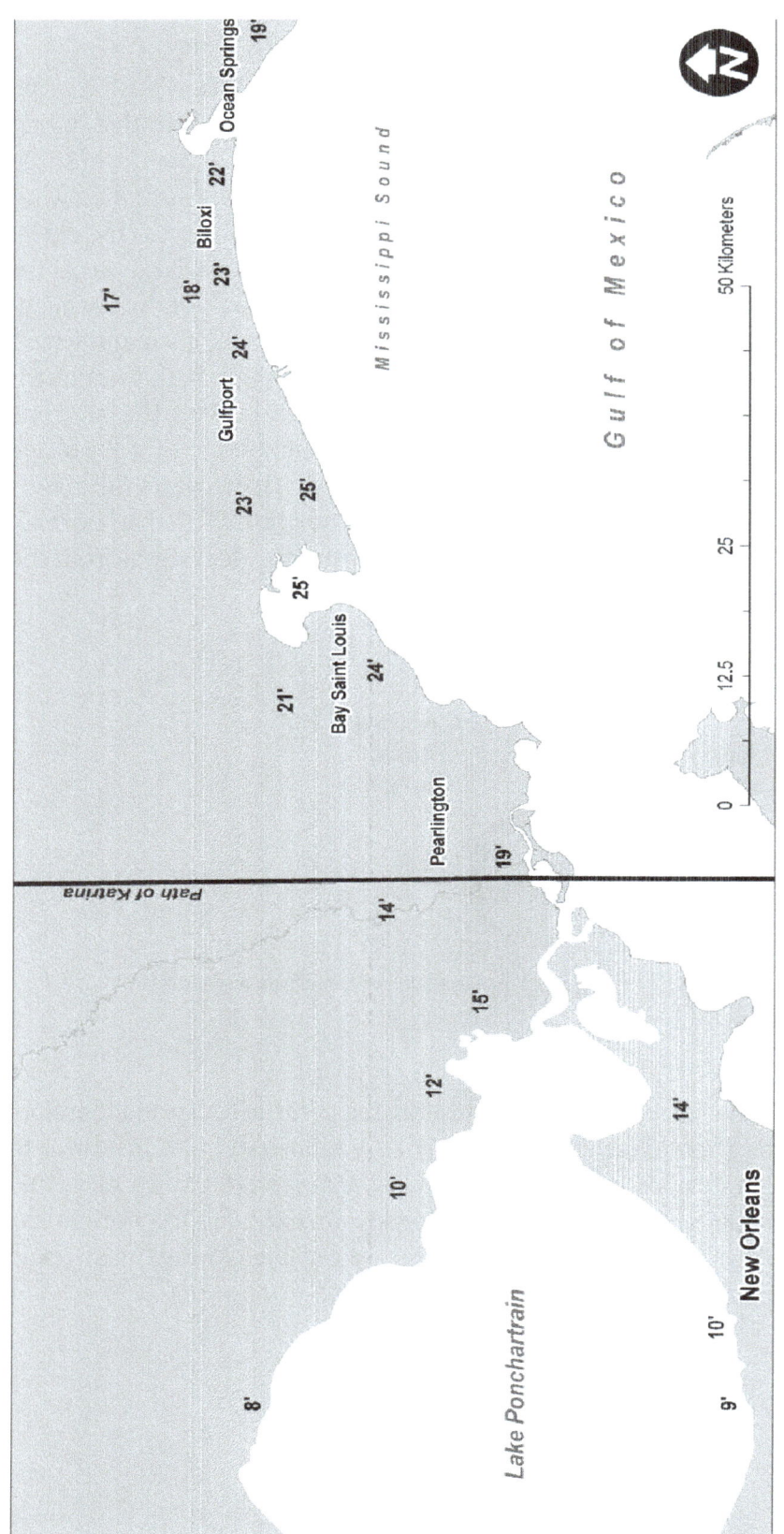

FIGURE 11.6 Maximum storm surge heights (feet above sea level) associated with Hurricane Katrina.

Use Figure 11.6 to answer the following questions.

1. When Hurricane Katrina made landfall just south of Pearlington, MS, in what direction should the surface winds have been blowing at the following locations?

   New Orleans, LA _____   Bay St. Louis _____

2. Where did the highest storm surge occur along the Gulf Coast?

   _____

3. How many kilometers away from the center of the storm did the highest storm surge occur?

   _____

4. Why was the storm surge so much greater around Gulfport than at the point of impact?

5. Why are the storm surge heights so much greater to the east of the point of impact compared to the west?

6. The city of New Orleans, Louisiana was devastated by the flood waters from Hurricane Katrina. Where did the flood waters come from? Hint: refer to question 1 about wind directions.

   _____

7. Bay St. Louis, Gulfport, Biloxi, and Ocean Springs, Mississippi sit on the Mississippi Sound. How did the shape of the sound impact the storm surge heights experienced at these locations?

Name _____ Section _____

LAB EXERCISE

Weather Analysis

12

In this exercise weather data at your location will be collected for a period of several days and then analyzed. The purpose of the exercise is to determine how major meteorological events can affect atmospheric conditions and to examine interrelationships among climatic parameters. For this exercise you are required to collect weather data for a period of up to 2 weeks before analysis begins.

Data Collection

Weather data will be collected once each day. The data will be collected at the same time every day. There are several sources that can be used, including radio or television weather reports. The data collection must be consistently from the same source.

The data will be recorded on the weather data log on the next page. If you miss collecting some data, do not use other sources or get the information from someone else. Omissions should be avoided. But missing minor subsets will not be critical. Data inconsistencies, caused by using different sources, may affect or invalidate your analysis.

Also, you should attempt to identify major changes in atmospheric conditions or meteorological events that occurred during the period of data collection, such as change in the positions of high pressure systems and cyclonic storms relative to your position, the passage of fronts, convectional precipitation, temperature inversions, etc. Record these conditions with the time and day they occurred.

Name _____

Group _____

Recording
Time _____

WEATHER DATA LOG

Date	Temp °C	Wind Direction	Wind Speed	Relative Humidity	Barometric Pressure	Precip (cm)	Cloud Cover (%)

After completing the above log, plot points for temperature and humidity on Figure 12.1, and barometric pressure and precipitation on Figure 12.2 for each day. Connect the points with lines for the different atmospheric variables shown. The lines will help you to visually interpret the change in conditions over time.

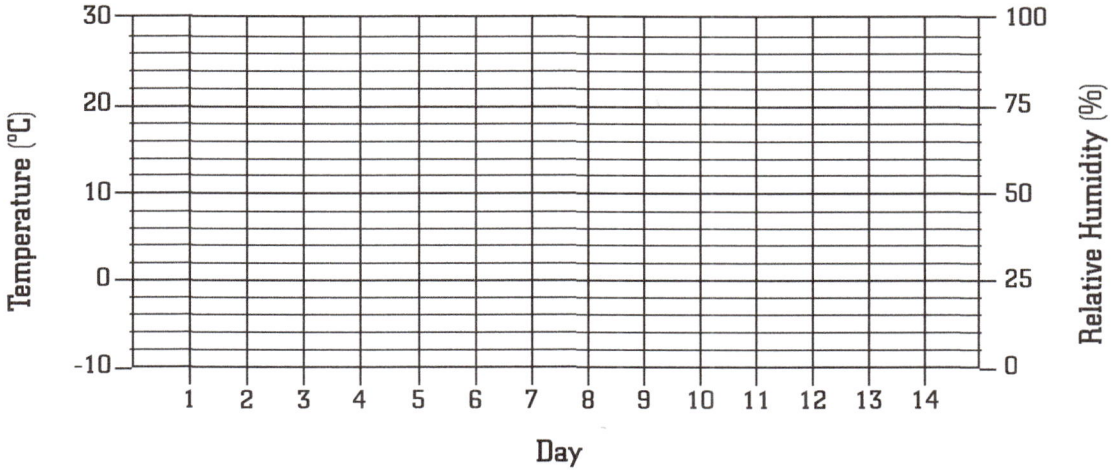

FIGURE 12.1 Temperature and humidity.

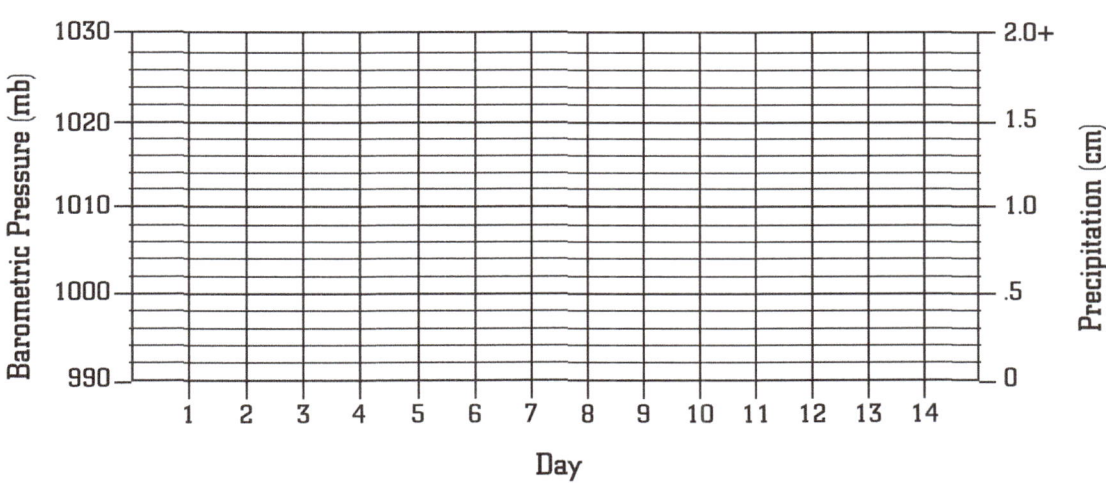

FIGURE 12.2 Barometric pressure and precipitation.

Use your Weather Data Log and the above graphs to answer the following questions.

1. What is the temperature range (max.–min.)? _____

    maximum temperature: _____

    minimum temperature: _____

2. What is the relative humidity range (max.–min.)? _____

    maximum RH: _____

    minimum RH: _____

3. Describe the relationship between temperature and humidity shown on Figure 12.1. Is there a direct or inverse relationship, and is this relationship strong or weak?

4. What is the barometric pressure range (max.–min.)? _____

    maximum pressure: _____

    minimum pressure: _____

5. It would not be unusual for no precipitation to occur over such a short period of time. Did precipitation occur during the period of data collection, and if so, describe any relationships between barometric pressure and precipitation (Figure 12.2).

Listed below are atmospheric phenomena that may or may not have affected your data. Compare the data that you have recorded in the Weather Data Log to the occurrence of these events (either that you noted or that is provided by the lab instructor). Describe and analyze these relationships. Include in these discussions when appropriate: changes in wind direction and speed, cloud cover and precipitation, temperature, humidity, and pressure.

1. Change in position of cyclonic storms:

2. Change in position of high pressure:

3. Passage of fronts:

4. _____

5. _____

Name_____ Section_____

LAB EXERCISE 13

Climate Classification

There are several climate classification systems used to describe similar climatic regions throughout the world. These classifications are based on different criteria, such as monthly or yearly precipitation and temperature, solar radiation, and length of the growing season. One of the most common climate classifications used by geographers and other earth scientists was developed by W. Koppen early this century and later revised by others.

Koppen's climate classification is shown in Figure 13.1. It is based on the global distribution of different types of vegetation. The type of vegetation in a region is determined primarily by climate, and boundaries for climatic regions and vegetation regions closely match. Koppen's climatic regions are designated by a series of letters. There are five basic climatic types. The general characteristics of each are listed below.

- A — Tropical Humid Climates
- B — Dry Climates
- C — Warm Temperate or Subtropical Climates
- D — Cool Temperate Climates
- E — Polar Climates

Subtypes are designated by second and third letters which indicate differences in temperature and precipitation. Monthly and annual temperature and precipitation averages are required to use this classification. An overview of the classification is shown in Table 13.1.

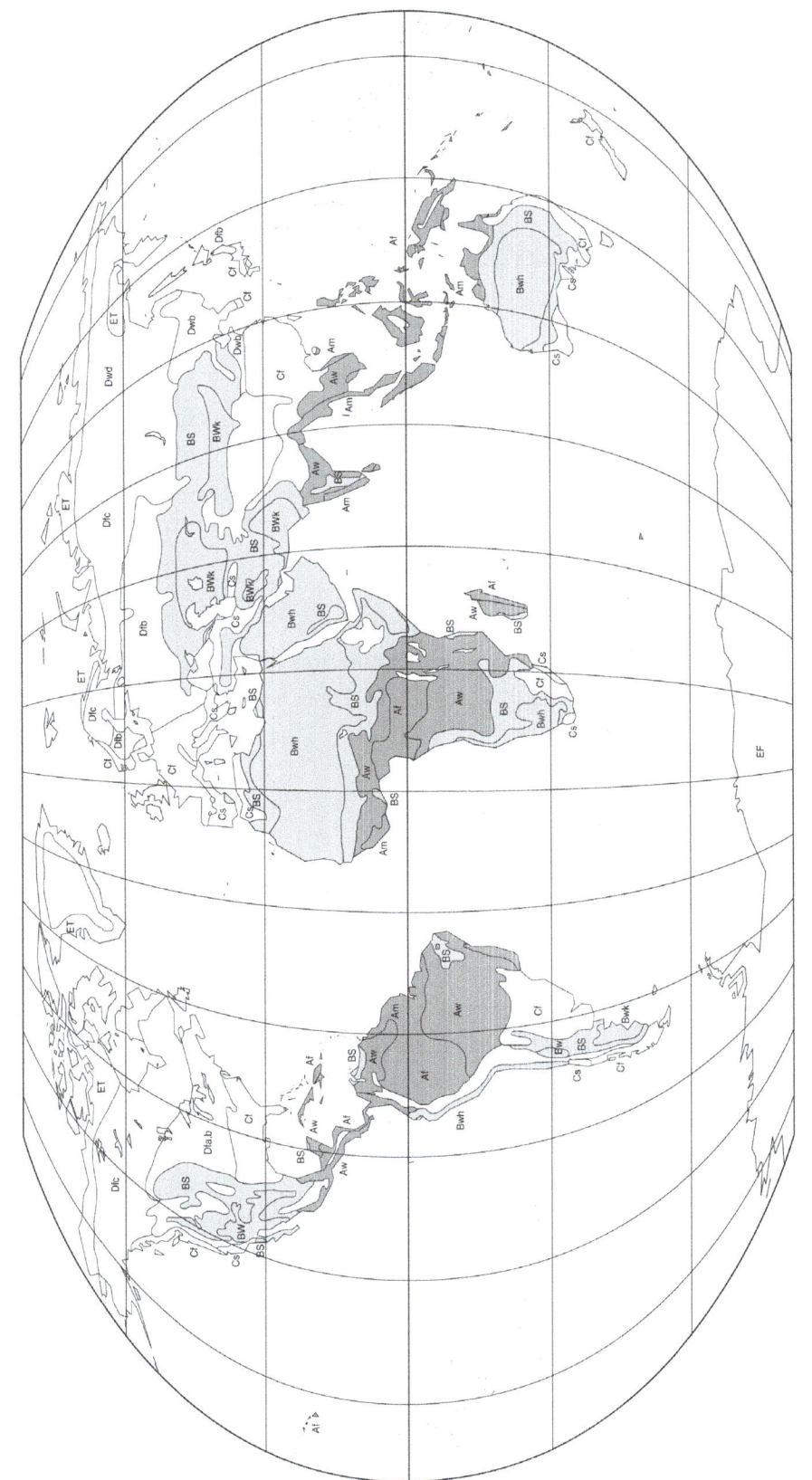

FIGURE 13.1 Map of Koppen's Climate Classification.

TABLE 13.1 KOPPEN CLIMATE CLASSIFICATION SYSTEM

Letter Symbol			Description and Criteria
1st	2nd	3rd	
A			**Tropical wet** (monthly temperature all greater than or equal to 18°C)
	f		Wet all year (every month receives 60mm or more of precipitation)
	m		Monsoon (short dry season—see Figure 13.3)
	w		Savanna (dry winter—see Figure 13.3)
B			**Dry** (potential evapo-transpiration exceeds precipitation—see Figure 13.2)
	W		Desert (see Figure 13.2)
	S		Steppe (see Figure 13.2)
		h	Tropical or hot (annual temperature is 18°C or greater)
		k	Mid-latitude or cool (annual temperature less than 18°C)
C			**Warm temperate** (average temperature of coldest month at least −3°C but less than 18°C; warmest month at least 10°C)
	w		Dry winter (driest winter month less than or equal to 1/10 of precipitation in wettest summer month)
	s		Dry summer (driest summer month less than or equal to 1/3 of precipitation in wettest winter month)
	f		Wet all year (all months receive at least 30mm precipitation)
		a	Hot summer (warmest month greater than or equal to 22°C)
		b	Warm long summer (at least 4 months have temperatures of 10°C or above but under 22°C)
		c	Short summer (1 to 3 months have temperatures of 10°C or above but under 22°C)
D			**Cool temperate** (coldest month less than −3°C; warmest month at least 10°C)
	w		same as C climate
	f		same as C climate
		a	same as C climate
		b	same as C climate
		c	same as C climate
		d	Severe winter (coldest month less than −38°C)
E			**Polar** (every month less than 10°C)
	T		Tundra with growing season (warmest month at least 0°C)
	F		Ice with no growing season (all months less than 0°C)

Procedure for Classifying Climates

To use Koppen's classification use the following procedure that tests for first letter classification in the following order: E, B, A, C, and then D. After establishing the first letter, determine the second letter classification, and then third letters for B, C, and D climates. Figure 13.2 is used to determine second letter B climates and to distinguish B climates from A, C, or D climates. Figure 13.3 is used to determine the second letter for A climates.

You must know if the climate station is in the northern or southern hemisphere. If you do not know the position of the climate station, the hemisphere can be determined by comparing monthly temperatures. If the coldest months are December, January, and February, it is in the northern hemisphere. If the coldest months are June, July, and August, it is in the southern hemisphere.

Note: T and P refer to average monthly values for temperature and precipitation.

1. If every T is less than 10°C, it is an E climate. If it is not an E climate, go to step 2.

    If every T is less than 0°C, it is an EF climate, if not it is an ET climate.

2. Use Figure 13.2 to determine if it is a B climate. If it is not a B climate go to step 3.

    a. Use Figure 13.2 to determine if it is a BW or BS climate.
    b. If annual temperature is less than 18°C, the third letter is k, either BWk or BSk. If annual temperature is greater than or equal to 18°C, the third letter is h, either BWh or BSh.

3. If every T is greater than or equal to 18°C, it is an A climate. If it is not an A climate go to step 4.

    a. If every P is greater than 60mm, it is Af.
    b. Use Figure 13.3 to determine if it is either Am or Aw.

4. If every T is greater than or equal to –3°C, it is a C climate. If it is not a C climate, go to step 5.

    Second letters

    a. If every P is greater than or equal to 30mm it is Cf.
    b. If the driest winter P is less than or equal to 1/10 of the wettest summer P it is Cw.
    c. If the driest summer P is less than or equal to 1/3 of the wettest winter P it is Cs. If not, it is Cf.

Third letters

d. If the warmest T is greater than or equal to 22°C, the third letter is **a**, either **Cfa**, **Cwa**, or **Csa**.
e. If at least 4 T are greater than or equal to 10°C, the third letter is **b**, either **Cfb**, **Cwb**, or **Csb**.
f. If fewer than 4 T are greater than or equal to 10°C, the third letter is **c**, either **Cfc** or **Cwc**. (**Csc** does not exist.)

5. It is a **D** climate.

Second letters

a. If every P is greater than or equal to 30mm it is **Df**.
b. If the driest winter P is less than or equal to 1/10 of the wettest summer P it is **Dw**, if not, it is **Df**.

Third letters

c. If the warmest T is greater than or equal to 22°C, the third letter is **a**, either **Dfa** or **Dwa**.
d. If at least 4 T are greater than or equal to 10°C, the third letter is **b**, either **Dfb** or **Dwb**.
e. If fewer than 4 T are greater than or equal to 10°C, the third letter is **c**, either **Dfc**, **Dwc**. (**Csc** does not exist.)
f. If the coldest month is less than −38°C, the third letter is **d**, either **Dfd** or **Dwd**.

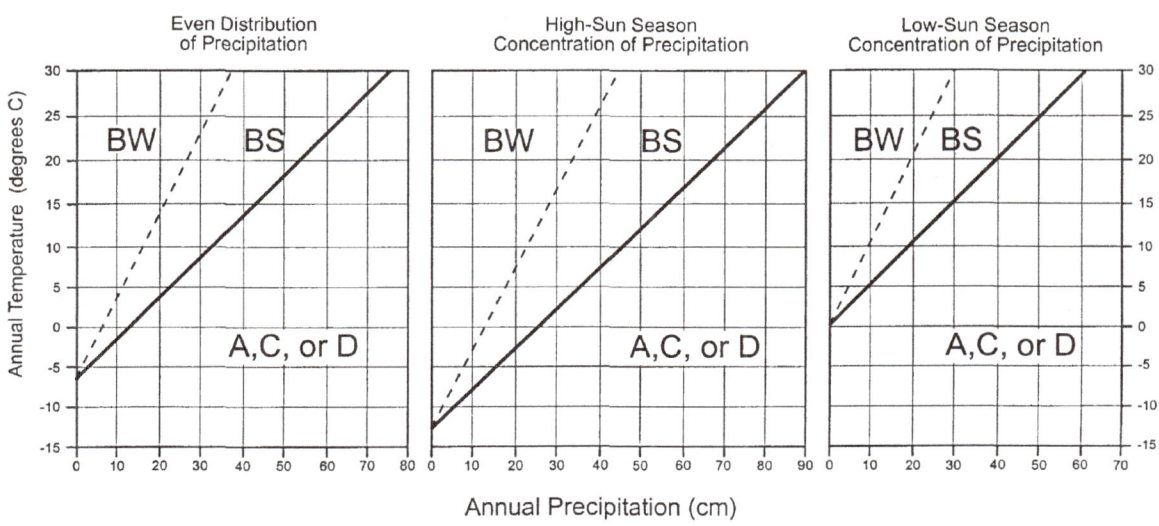

FIGURE 13.2 Determine second letter B climates and distinguish B climates from A, C, or D climates.

Figure 13.2 is used to determine second letter B climates and to distinguish B climates from A, C, or D climates. To use these diagrams, follow these procedures:

1. Determine if the climate station is in the northern or southern hemisphere by comparing monthly temperatures.
2. Determine whether most of the precipitation occurs in the summer (high sun season), winter, (low sun season), or is evenly distributed throughout the year. For the purpose of this classification, the year is divided into two six month seasons, April–September and October–March.
3. If 70 percent or more of the annual total precipitation is in one of these six month periods, that season is said to have a precipitation concentration.
4. Use the appropriate one of these three diagrams and plot annual total precipitation and annual temperature values. If a point is plotted on the right side of the graph, it is not a dry climate, and is either an A, C, or D climate. If the point is plotted on the left side of the graph, it is a B climate.
5. If the climate station is classified as a B climate, use the graph to determine if it is a BS (desert steppe) or BW (true desert) which is drier.

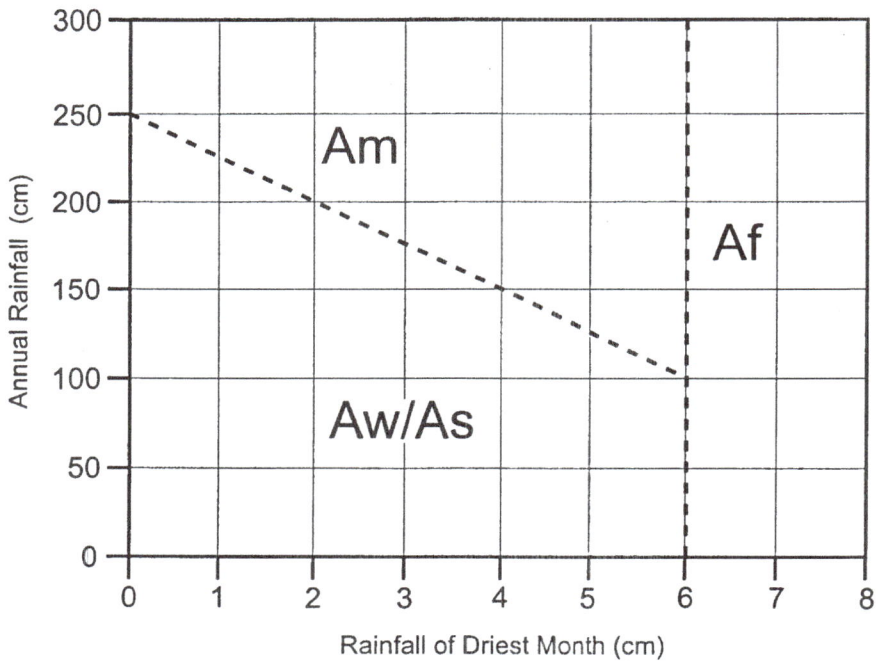

FIGURE 13.3 Determine the second letter for A climates.

Determine the Koppen classification for the following climate stations:

Note: Temperature is degrees C, precipitation is in mm. **Yr** values for temperature is an average of all the months. **Yr** values for precipitation are yearly totals.

Atlanta, Georgia, USA
Location: 34°N; 84°W. Elevation: 308m

	Jan	Feb	Mar	Apr	May	Jun	Jul	Aug	Sep	Oct	Nov	Dec	Yr
T	6	7	11	16	21	25	26	25	23	17	11	7	16
P	143	136	148	131	114	105	144	110	84	92	103	121	1437

Climate classification: _____

Cape Town, South Africa
Location: 34°S; 19°E. Elevation: 12m

	Jan	Feb	Mar	Apr	May	Jun	Jul	Aug	Sep	Oct	Nov	Dec	Yr
T	21	21	20	17	15	13	12	13	14	16	18	20	17
P	16	15	21	50	92	105	91	82	54	40	24	19	612

Climate classification: _____

Chicago, Illinois, USA
Location: 42°N; 88°W. Elevation: 205m

	Jan	Feb	Mar	Apr	May	Jun	Jul	Aug	Sep	Oct	Nov	Dec	Yr
T	−4	−2	4	10	16	21	24	24	20	14	6	−1	11.0
P	48	40	73	91	99	93	91	76	65	69	61	61	905

Climate classification: _____

Beijing, China
Location: 40°N; 116°E. Elevation: 54m

	Jan	Feb	Mar	Apr	May	Jun	Jul	Aug	Sep	Oct	Nov	Dec	Yr
T	−4	−1	4	13	20	24	26	25	20	13	4	−3	11
P	3	4	8	18	33	78	224	170	58	18	9	2	635

Climate classification: _____

Phoenix, Arizona, USA
Location: 33°N; 112°W. Elevation: 337m

	Jan	Feb	Mar	Apr	May	Jun	Jul	Aug	Sep	Oct	Nov	Dec	Yr
T	11	13	16	20	25	30	33	32	28	22	15	11	21
P	16	20	19	6	3	1	16	29	22	14	14	27	189

Climate classification: _____

Ho Chi Min City, Vietnam
Location: 11°N; 107°E. Elevation: 19m

	Jan	Feb	Mar	Apr	May	Jun	Jul	Aug	Sep	Oct	Nov	Dec	Yr
T	27	27	28	29	29	28	27	27	27	27	26	26	27
P	14	4	9	51	213	309	295	271	342	261	119	47	1903

Climate classification: _____

Fairbanks, Alaska, USA
Location: 65°N; 148°W. Elevation: 189m

	Jan	Feb	Mar	Apr	May	Jun	Jul	Aug	Sep	Oct	Nov	Dec	Yr
T	−21	−18	−10	−1	9	15	16	13	7	−4	−15	−20	−3
P	15	13	10	7	17	45	56	53	32	22	19	21	310

Climate classification: _____

Paris, France
Location: 49°N; 2°E. Elevation: 66m

	Jan	Feb	Mar	Apr	May	Jun	Jul	Aug	Sep	Oct	Nov	Dec	Yr
T	3	4	6	10	14	17	19	18	15	11	6	3	11
P	46	39	41	44	56	57	57	55	53	57	54	49	607

Climate classification: _____

Tropical Climates (A)

Tropical climates are warm and humid and occur mostly within 20° of the equator. These regions do not have a winter season and the annual temperature range is small. Annual precipitation in most tropical regions is high. The year-round growing season and high precipitation support forest or mixed forest-grassland vegetation.

Most rainfall is caused by convection related to the Intertropical Convergence Zone (ITCZ). Migration of the ITCZ with changing declination of the sun causes seasonal variability in precipitation. Precipitation is also caused by (1) convection related to low pressure troughs in the region of the trade winds, referred to as *easterly waves*, and (2) tropical cyclones.

There are three subtypes of Tropical climates. First, Tropical Wet (Af) climates have year-round precipitation related to the ITCZ, and therefore, no dry season. Tropical Savannas (Aw) are under alternating influence of the ITCZ (wet season) and subtropical high pressure (dry season). Tropical Monsoons (Am) occur along tropical coasts backed by highlands. Like the Aw climates, these regions have alternating wet summer and dry winter seasons. However, Tropical Monsoons have a comparatively long and heavy rainy wet summer and a short dry winter.

Bangui, Central African Republic
Location: 4°N; 19°E. Elevation: 365m

	Jan	Feb	Mar	Apr	May	Jun	Jul	Aug	Sep	Oct	Nov	Dec	Yr
T	25	26	27	27	26	25	24	24	25	25	25	25	25
P	19	38	106	132	163	142	180	224	190	201	92	28	1507

Climate classification: _____

Kinshasa, Democratic Republic of the Congo
Location: 4°S; 15°E. Elevation: 309m

	Jan	Feb	Mar	Apr	May	Jun	Jul	Aug	Sep	Oct	Nov	Dec	Yr
T	25	25	25	25	25	22	22	22	25	25	25	25	25
P	138	148	184	219	148	5	3	4	40	133	235	156	1406

Climate classification: _____

Mergui, Myanmar
Location: 12°N; 99°E. Elevation: 2m

	Jan	Feb	Mar	Apr	May	Jun	Jul	Aug	Sep	Oct	Nov	Dec	Yr
T	26	27	28	28	28	27	26	26	26	27	27	26	27
P	22	50	77	131	438	772	808	768	619	312	88	19	4092

Climate classification: _____

Manaus, Brazil
Location: 3°S; 60°W. Elevation: 65m

	Jan	Feb	Mar	Apr	May	Jun	Jul	Aug	Sep	Oct	Nov	Dec	Yr
T	26	26	26	26	26	26	26	27	27	28	27	27	27
P	264	262	298	282	204	103	67	46	63	111	161	220	2088

Climate classification: _____

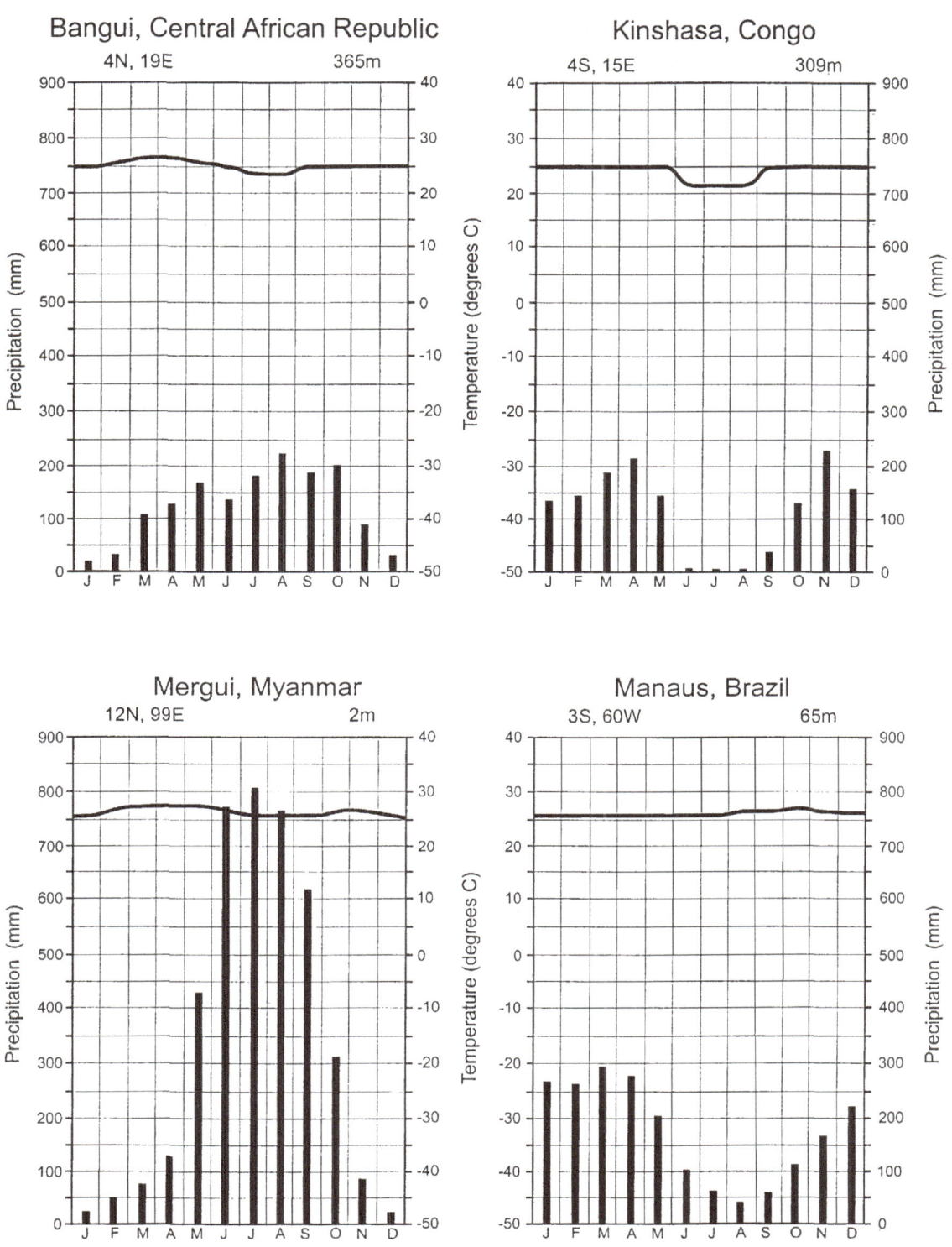

FIGURE 13.4 Tropical climates.

1. What is the annual temperature range for each location?

   Bangui, Central Africa Republic: _____

   Kinshasa, Democratic Republic of the Congo: _____

   Mergui, Mayanmar: _____

   Manaus, Brazil: _____

2. The temperature range for the tropics is small compared to higher latitudes. Why?

3. Do these locations have higher precipitation in the summer or winter? Explain this climatic pattern.

Dry Climates (B)

Desert and Steppe climate regions are defined by lack of water. They cover about 30 percent of the world's land area. These regions occur from low to high latitudes, and temperatures range from cold to hot. No permanent streams originate in these regions. Although permanent streams with headwaters in more humid regions will flow through these dry regions. True deserts (BW) support sparse or no vegetation except along streams. Steppe regions (BS) have somewhat more precipitation, although still too dry to support trees. Dry climates are the result of (1) rainshadow effect on the lee side of mountain ranges, (2) dominance of subtropical high pressure, (3) remoteness from moisture sources in the interior of large northern hemisphere continents, and (4) cold ocean current off the west coast of continents.

The daily temperature range in dry climates is often very high. Typically, there is little cloud cover and humidity is low. Also, there is little evaporative cooling of the surface. Daytime surface temperature can be very high. In contrast, temperatures can fall rapidly after sunset because of the rapid loss of longwave radiation.

Cairo, Egypt
Location: 30°N; 31°E. Elevation: 32m

	Jan	Feb	Mar	Apr	May	Jun	Jul	Aug	Sep	Oct	Nov	Dec	Yr
T	13.8	15.2	17.4	21.4	24.7	27.3	27.9	27.9	26.3	23.7	19.1	15.1	21.7
P	5	4	4	2	1	0	0	0	0	1	3	6	26

Climate classification: _____

Alice Springs, Australia
Location: 24°S; 134°E. Elevation: 537m

	Jan	Feb	Mar	Apr	May	Jun	Jul	Aug	Sep	Oct	Nov	Dec	Yr
T	29	28	25	20	16	12	12	14	18	23	26	28	21
P	41	42	35	17	17	17	12	10	9	20	25	37	281

Climate classification: _____

Iquiqeu, Chile
Location: 21°S; 70°W. Elevation: 10m

	Jan	Feb	Mar	Apr	May	Jun	Jul	Aug	Sep	Oct	Nov	Dec	Yr
T	21	21	20	18	17	16	16	15	16	17	18	20	18
P	0	0	0	0	0	1	1	0	0	0	0	0	2

Climate classification: _____

Death Valley, California, USA
Location: 36°N; 117°W. Elevation: –59m

	Jan	Feb	Mar	Apr	May	Jun	Jul	Aug	Sep	Oct	Nov	Dec	Yr
T	11	15	19	24	30	35	38	37	32	25	17	10	24
P	8	12	9	4	2	1	3	3	4	3	6	4	60

Climate classification: _____

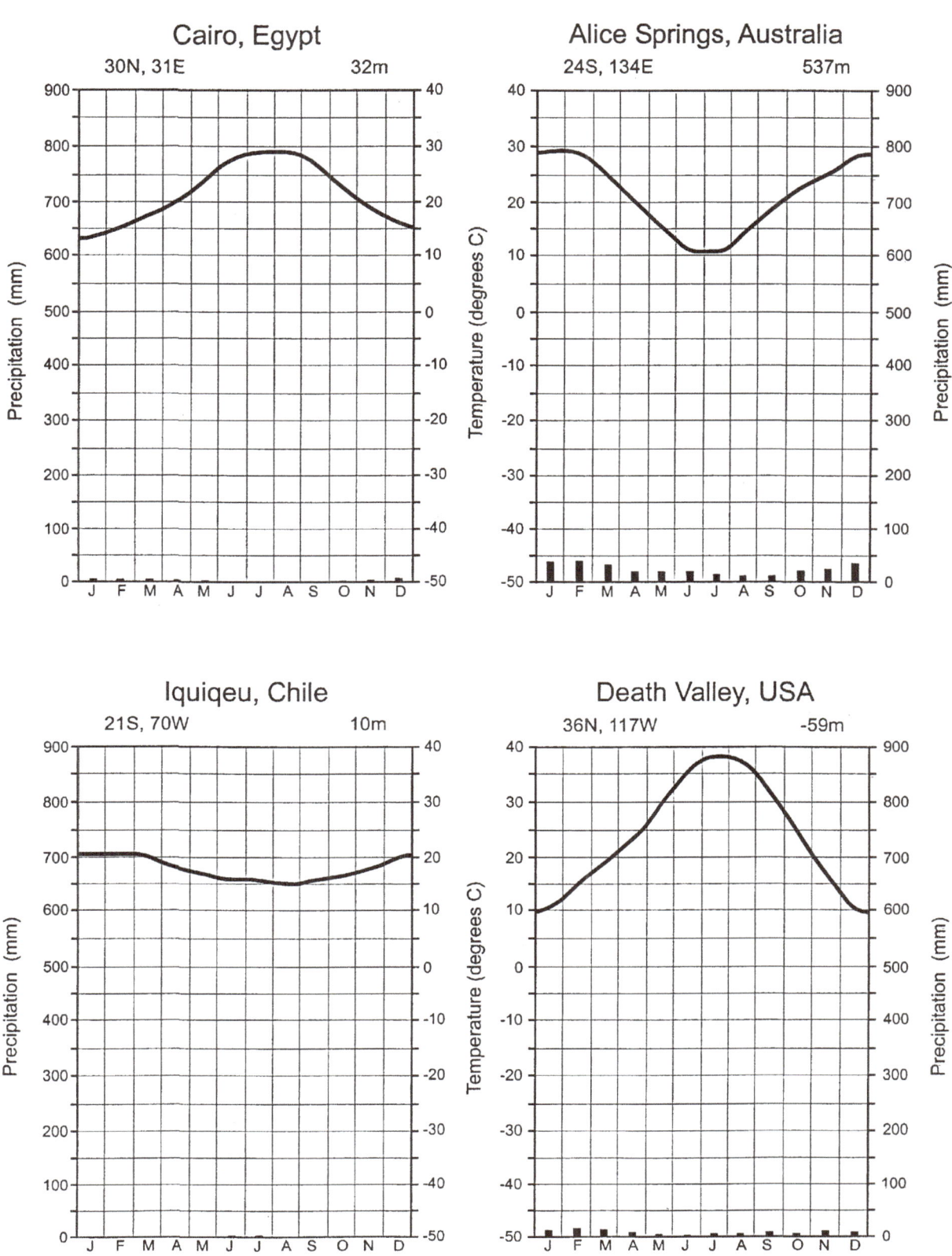

FIGURE 13.5 Dry climates.

1. What is the annual temperature range for each location?

   Cairo, Egypt: _____

   Alice Springs, Australia: _____

   Iquiqeu, Chile: _____

   Death Valley, California, USA: _____

2. What is the relationship between latitude and annual temperature range?

3. What are the dominant climatic controls causing dry conditions in each of these locations?

   Cairo, Egypt:

   Alice Springs, Australia:

   Iquiqeu, Chile: **Note:** This location is on the west coast of South America.

   Death Valley, California, USA:

Mid-Latitude Climates (C and D)

Mid-latitude climates are regions of conflicting air masses with alternating incursions of polar and tropical air. They are characterized by distinct seasonal changes in temperature and precipitation. Winter temperatures range from bitter cold in the interior of Asia and North America to mild along the coasts at lower latitudes. Summer temperatures range from hot to mild.

The dominant precipitation mechanisms are cyclonic storms and fronts. These are most common during the winter and spring. Convection is a common cause of precipitation during the summer, especially on the southeastern parts of continents because of high summer-time humidity in these regions.

Warm Temperate climates (C) occur mostly on the western and southeastern portions of continents. Cold continental climates (D) cover much of the high and mid-latitudes of the large continents of the northern hemisphere. Cold continental climates do not occur in the southern hemisphere because of the absence of large land masses in the mid-latitudes.

London, England
Location: 52°N; 0°W. Elevation: 24m

	Jan	Feb	Mar	Apr	May	Jun	Jul	Aug	Sep	Oct	Nov	Dec	Yr
T	5	5	7	9	13	16	18	18	15	12	8	6	11
P	62	36	50	43	45	46	46	44	43	73	45	59	588

Climate classification: _____

Athens, Greece
Location: 38°N; 24°E. Elevation: 28m

	Jan	Feb	Mar	Apr	May	Jun	Jul	Aug	Sep	Oct	Nov	Dec	Yr
T	10	11	12	16	21	25	28	28	24	19	15	12	19
P	48	41	41	23	18	7	5	8	10	53	55	62	371

Climate classification: _____

Fuzhou, China
Location: 26°N; 119°E. Elevation: 152m

	Jan	Feb	Mar	Apr	May	Jun	Jul	Aug	Sep	Oct	Nov	Dec	Yr
T	11	11	14	18	22	26	29	29	26	22	18	14	20
P	47	87	120	139	179	210	138	172	165	47	40	38	1389

Climate classification: _____

Irkutsk, Russia
Location: 64°N; 100°E. Elevation: 650m

	Jan	Feb	Mar	Apr	May	Jun	Jul	Aug	Sep	Oct	Nov	Dec	Yr
T	−36	−32	−19	−7	3	12	16	12	5	−7	−26	−33	−9
P	13	10	9	13	25	51	62	54	36	25	20	16	335

Climate classification: _____

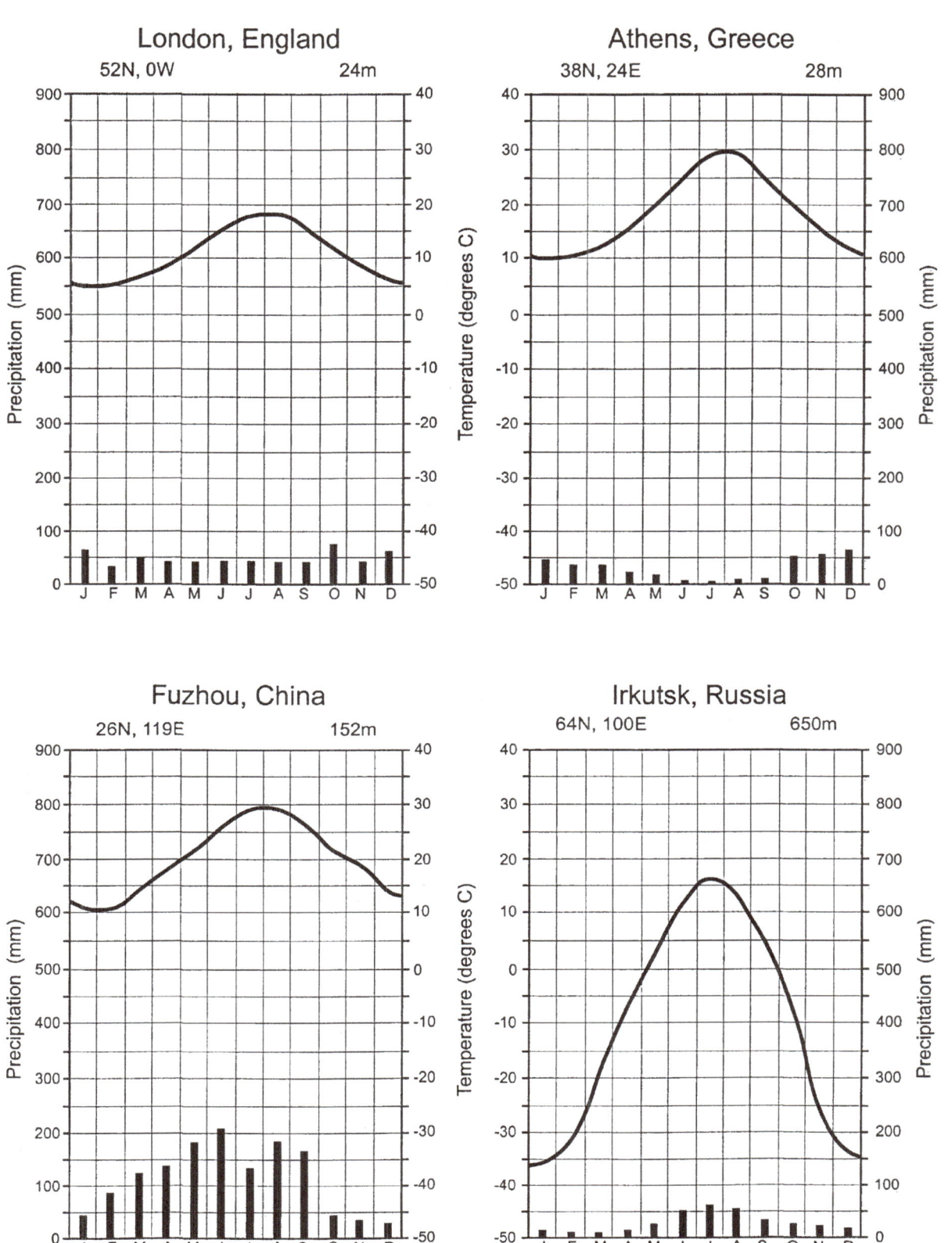

FIGURE 13.6 Mid-latitude climates.

1. What is the annual temperature range for each location?

   London, England: _____

   Athens, Greece: _____

   Fuzhou, China: _____

   Irkutsk, Russia: _____

2. London, England is located 52°N, a higher latitude than all areas in the continental U.S. However, London has mild winter temperatures and a low annual temperature range. Explain why this location has moderate temperatures.

3. Irkutsk, Russia is located in central Asia. Explain why this location has an extremely high annual temperature range.

4. What climatic control explains the distinct wet winter and dry summer occurring in Athens, Greece?

5. Mid-latitude cyclones and fronts occur more frequently during the winter and these are the dominant precipitation mechanisms for most of the mid-latitudes. Why does Fuzhou, China have a wet summer season?

Name _____ Section _____

Topographic Map and Aerial Photograph Interpretation

LAB EXERCISE **14**

Topographic maps are among the most useful tools for earth scientists. They show physical features of the earth's surface (such as elevation, streams, glaciers, vegetation), man-made features (buildings, roads, mines), and geographical grid systems that are used to determine location. The purpose of this exercise is for you to become familiar with the features of these maps, including map symbols, scale, and contour lines. It is important that you are able to interpret topographic maps since they will be used extensively with the later exercises in this lab manual.

Topographic maps use a standard set of symbols to identify both natural and human features of the landscape. A legend of these symbols is placed in the back of this text. Since you will be using these maps extensively in the following exercises you need to be familiar with map symbols.

Sections of ten topographic maps are included in this manual to illustrate a variety of landscapes. The map scale and contour intervals are noted on the back of each. These maps are only portions of the printed originals and do not include information included in the margins (such as publication date, magnetic declination, grid coordinates, and names of adjoining maps).

A complete quadrangle should be available for students at the beginning of this exercise. The map sections that are included will be sufficient for the remainder of exercises in this manual. But the lab instructor may prefer to use maps that illustrate features in your area.

A topographic map will be provided by your lab instructor.

1. The map _____
   (name of quadrangle)

2. Geographic grid coordinates at the lower right corner:

   _____

3. Date of map: _____

4. Scale (Relative Fraction): _____

5. Contour interval: _____ feet

6. What quadrangle joins this on the south? _____

7. What information is represented by these colors?

   Blue: _____

   Green: _____

   Red: _____

Map Scale

One of the most important features of a map is its scale. Map scale is the ratio between the distance on the map and the corresponding distance on the ground. The ability to work with map scale is necessary for proper interpretation.

The most commonly used scale is **Representative Fraction** (R.F.) which is usually shown as a ratio:

$$\text{distance on map:distance on ground}$$

An R.F. of 1:12,000 means that one inch on the map equals 12,000 inches on the ground. Of course, different units of measurement (such as cm or feet) can be used, but you must use the same units on both sides of the ratio.

Conversion of Units

You can rapidly convert from one scale unit to another. Suppose you want to convert 12,000 inches to miles. Set up the problem so that you multiply the number of inches (12,000) by 1 mile divided by its equivalent in inches. There are 63,360 inches per mile. So:

$$12{,}000 \text{ inches} \ * \ \frac{1 \text{ mile}}{63{,}360 \text{ inches}}$$

Inches cancel on the top and bottom so that only the mile unit remains. Then multiply:

$$12{,}000 \text{ inches} \ * \ \frac{1 \text{ mile}}{63{,}360 \text{ inches}} \ = \ \frac{12{,}000 \text{ mile}}{63{,}360} \ = \ 0.189 \text{ miles}$$

Examples

1. Convert 800 inches to feet:

$$800 \text{ inches} \ * \ \frac{1 \text{ foot}}{12 \text{ inches}} \ = \ \frac{800 \text{ feet}}{12} \ = \ 66.6 \text{ feet}$$

2. Convert 1600 feet to miles:

$$1600 \text{ feet} \ * \ \frac{1 \text{ mile}}{5280 \text{ feet}} \ = \ \frac{1600 \text{ miles}}{5280} \ = \ 0.30 \text{ miles}$$

You can also use this method to easily convert between English and metric units:

3. Convert 1334 centimeters to inches:

$$1334 \text{ cm} * \frac{1 \text{ inch}}{2.54 \text{ cm}} = \frac{1334 \text{ inches}}{2.54} = 525.2 \text{ inches}$$

4. Convert 22.3 miles to kilometers:

$$22.3 \text{ miles} * \frac{1 \text{ km}}{0.621 \text{ miles}} = \frac{22.3 \text{ km}}{0.621} = 35.9 \text{ km}$$

English measurements can be confusing because the corresponding units do not have consistent conversion ratios (12 inches = 1 foot, 5280 feet = 1 mile). This is one of the major reasons the U.S. is switching to the metric system, which is used by most of the world's industrialized countries. Metric units can be divided into even numbers of 10, 100, or 1000.

1 centimeter (cm)	=	10 millimeters (mm)
1 meter (m)	=	100 centimeters
1 kilometer (km)	=	1000 meters

Example

With an R.F. of 1:12,000, one centimeter on the map = 12,000 centimeters on the ground. The distance on the ground (in centimeters) can be converted to meters or kilometers:

$$12{,}000 \text{ cm} * \frac{1 \text{ m}}{100 \text{ cm}} = \frac{12{,}000 \text{ m}}{100} = 120 \text{ m}$$

$$12{,}000 \text{ cm} * \frac{1 \text{ km}}{100{,}000 \text{ cm}} = \frac{12{,}000 \text{ km}}{100{,}000} = 0.12 \text{ km}$$

The Seagrove map shows an area of central North Carolina. The map has a Relative Fraction (R.F.) of 1:24,000. What is the distance on the ground between the Fairview Church, southeast of the town of Seagrove and the Fulp Church, north of Seagrove?

1. First, measure the distance on the map between these two points.

   _____ (cm or inches)

2. Multiply this value by the denominator of the R.F. to determine the distance on the ground.

   _____

3. Convert this distance to either feet or meters, depending on if you are using English or metric units.

   _____ (feet or meters)

Enter this value below and convert to the other units shown.

_____ feet

_____ miles

_____ meters

_____ kilometers

4. What is the distance across the Seagrove map? _____

Refer to the maps located in different chapters for the following questions.

5. **Heidelberg, Kentucky** (Lab Exercise 21)

    R.F.: _____

    Width of the Kentucky River at the mouth of Cave Branch:

    _____

6. **Cape Solitude, Arizona** (Lab Exercise 23)

    R.F.: _____

    Distance from Sixtymile Rapids to the mouth of the Little Colorado River:

    _____

7. **Fishhook Lake, Mississippi** (Lab Exercise 23)

    R.F.: _____

    Shortest distance between Black Lake and the Tallahatchie River:

    _____

8. **Stovepipe Wells, California** (Lab Exercise 24)

    R.F.: _____

    State the verbal scale of the map:

    a. inches to miles

        Scale: _____ inch = _____ mile

    b. centimeters to kilometers

        Scale: _____ cm = _____ km

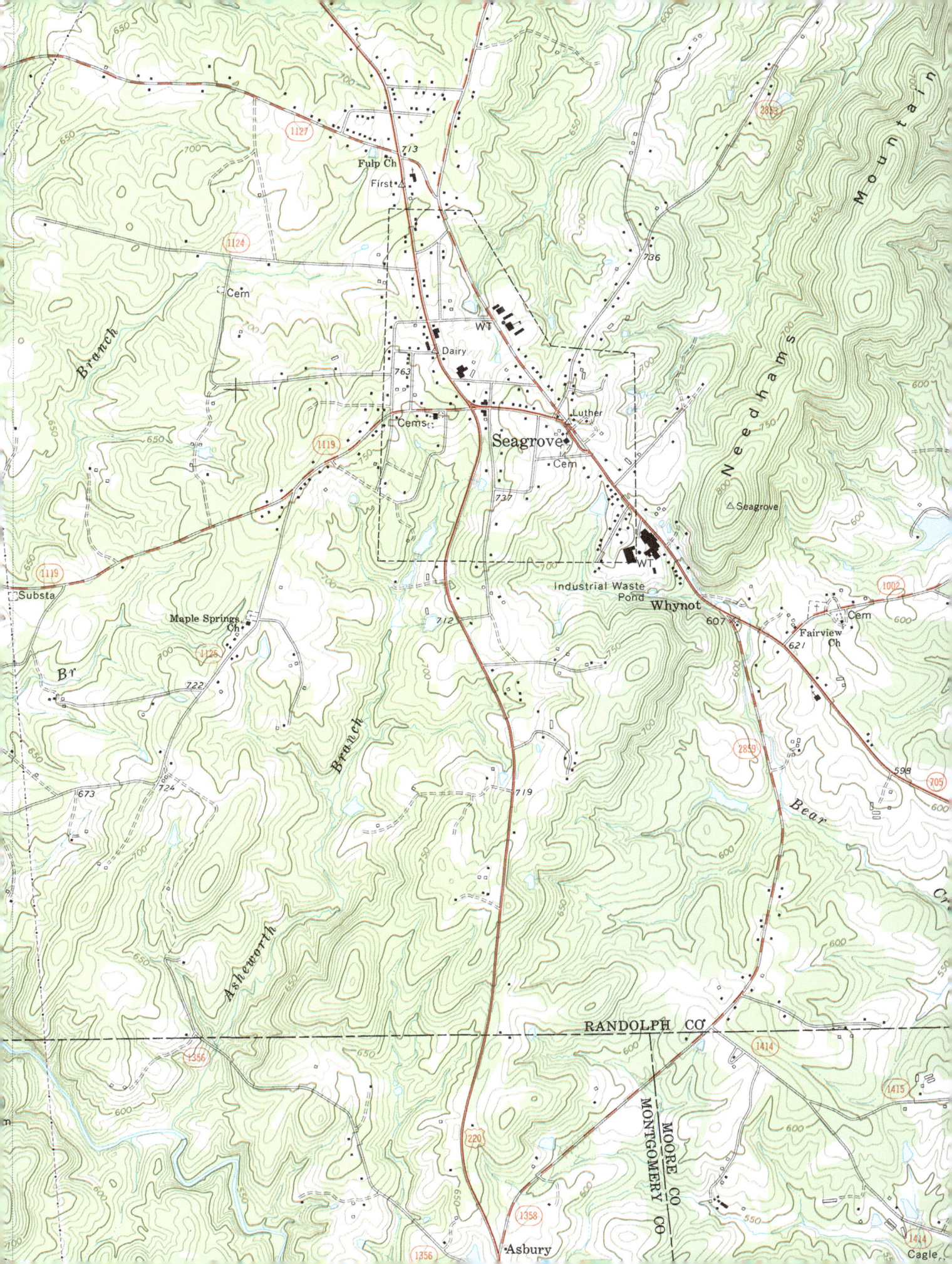

SEAGROVE, NORTH CAROLINA

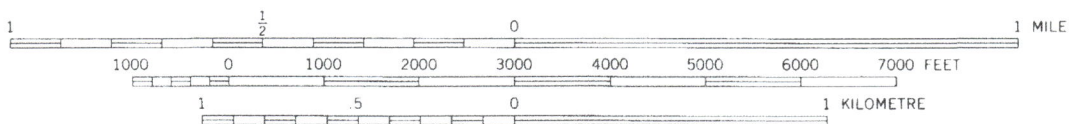

SCALE 1:24,000

Contour Interval 10 Feet

Calculating Area

The R.F. can be used to determine area. For example, if the R.F. = 1:24,000, one inch on the map equals 24,000 inches on the ground, or 2,000 feet. Therefore, 1 square inch on the map or photograph (1 × 1), the area on the ground would be calculated as:

$$2000 \times 2000 = 4,000,000 \text{ feet}^2$$

1. Assume a map has a scale of 1: _____. If the area on the topographic map is _____ × _____ (in. or cm), determine the corresponding area on the ground.

   _____ feet^2

   _____ mile^2

   _____ meters^2

   _____ km^2

2. Assume a map has a scale of 1: _____. If the area on the topographic map is _____ × _____ (in. or cm), determine the corresponding area on the ground.

   _____ feet^2

   _____ mile^2

   _____ meters^2

   _____ km^2

Circular areas can also be easily calculated. Refer to Figure 14.1 shown below.

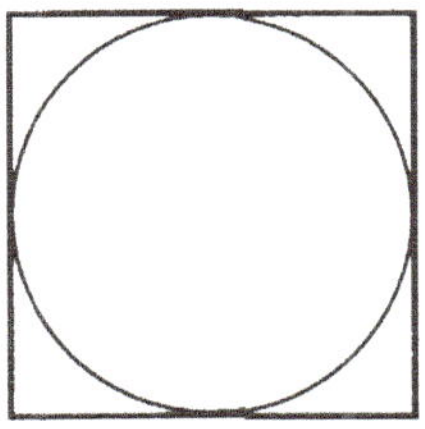

FIGURE 14.1 Circular area.

3. To determine the area of a circle in this problem, use the following R.F.: _____.

   The side of the square is _____ inches, which is also the diameter of the circle.

   The diameter of the circle on the ground = _____ feet

   The area of the circle = 3.14 * radius^2

   The area of the circle = _____ feet^2

4. Refer to the Mt. Capulin aerial photograph later in this exercise. The R.F. is 1:20,000.

   What is area of the cone? _____

   What is the area of the crater? _____

Topographic Contours

A contour line is a line on a map that connects points having equal elevation, or distance above (or below) sea level. These are among the most useful features of maps because they show how the landscape varies and can be used to identify different types of landforms.

The elevational distance between two contour lines is the contour interval. The contour interval used depends primarily on the map scale and the relief of the topography. Intervals of 5, 10, or 20 feet are common.

Contour Line Characteristics

1. A contour line represents only one elevation.
2. Contour lines do not divide or split.
3. Contour lines do not end. They will reconnect, but may run off maps that represent only a small portion of the landscape.
4. Contours are closer together on steep slopes than gentle slopes, but they never cross one another. On vertical slopes contour lines will overlap.
5. Contour lines bend upstream into an inverted "V" when crossing streams, and bend downslope when crossing ridges.
6. The contour interval is constant unless otherwise noted.
7. Hatchured contours indicate a depression. The hatchured lines have small tick marks on the downhill side.

Figure 14.2 shows contour lines and the corresponding landscape profile.

The south and western part of the Platte City map includes a portion of the Missouri River and Platte River floodplain. The wide spacing of contour lines in these areas indicate that these surfaces are nearly flat. In contrast, the upland areas adjacent to the floodplain have a more rugged terrain. The closely spaced contour lines on this section of the map indicate steeper slopes.

Answer the following using the Platte City map:

1. Contour interval: _____

2. Elevation of Green Cemetery is _____ ft

3. Elevation of _____ is _____ ft

4. Elevation of _____ is _____ ft

5. What is the location of the highest point on the map? (name or coordinates)

   What is the elevation of that point? _____

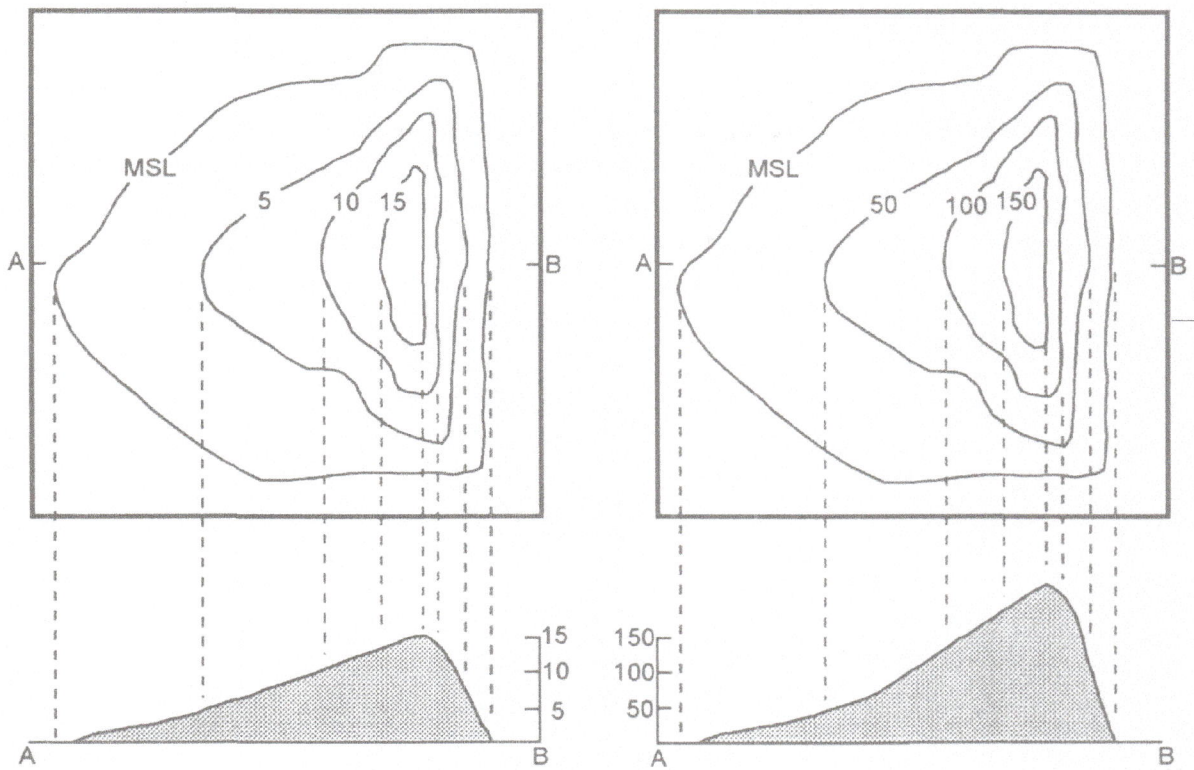

FIGURE 14.2 Landscape profile and corresponding contours.

6. What is the location of the lowest point on the map? (name or coordinates)

   What is the elevation of that point? _____

7. **Relief** is the difference between the highest and lowest points of a specific area. What is the relief for the area shown on this map?

   _____

8. What area of the map has the steepest slopes? (name or coordinates)

9. In which direction does the Platte River flow?

10. In what direction does the bluff on which the town of Farley occurs face?

    _____

11. Why do you think the town of Farley was built on the bluff rather than in the floodplain?

Drawing Contour Lines

On Figure 14.4 is a map with elevational points. The 170 foot contour has already been drawn. Draw the remaining contour lines at intervals of 10 feet. Be sure that all points with a lower elevation than the contour line you are drawing are on the downhill side of the line. Use a pencil and draw lightly until you are sure the contour positions are correct, then darken the lines. Be sure the inverted "V"s coincide with the stream channels.

Slope Gradient

The slope of the land surface can easily be determined on contour maps. The gradient, or slope angle, is often expressed as a ratio:

$$\frac{\text{rise}}{\text{run}}$$

where:

rise = the elevational difference between 2 points
run = the horizontal distance between them.

For example, on the cross-section of a hillslope shown in Figure 14.3, point A is upslope of point B.

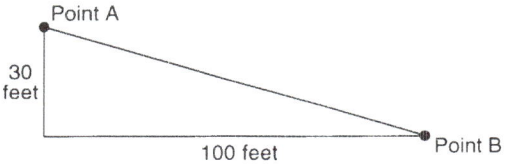

FIGURE 14.3 Cross-section of a hillslope.

Difference in elevation between points A and B: 30 feet
Horizontal distance between points A and B: 100 feet

The slope can be expressed in one of the following ways:

a. Percent grade: 30%

$$\frac{30}{100} = .30 = 30\%$$

b. Feet per mile: 1584 feet per mile

$$\frac{30}{100} = \frac{x}{5280}$$

$$100x = 158{,}400$$
$$x = 1584 \text{ feet per mile}$$

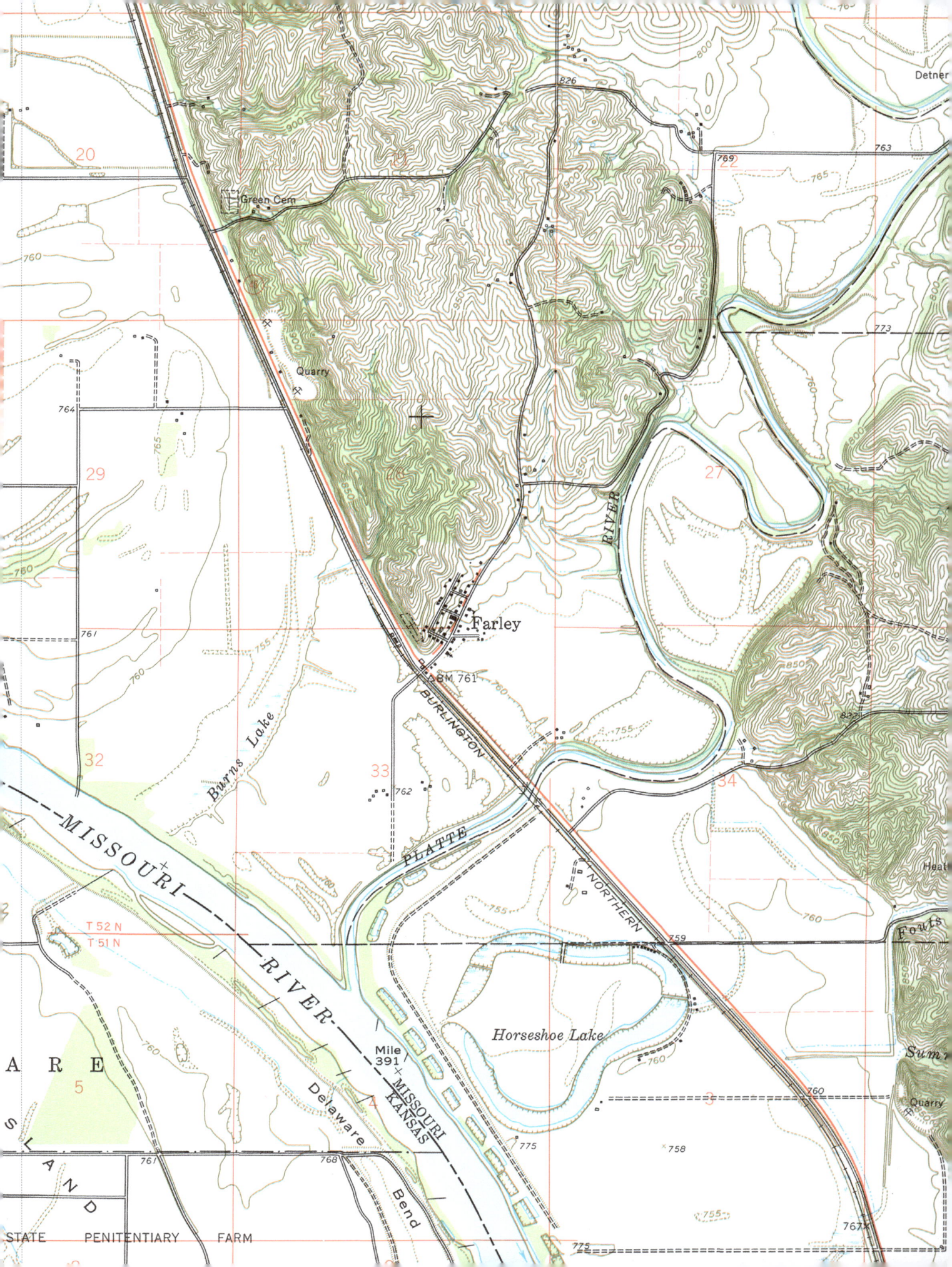

PLATTE CITY, MISSOURI–KANSAS

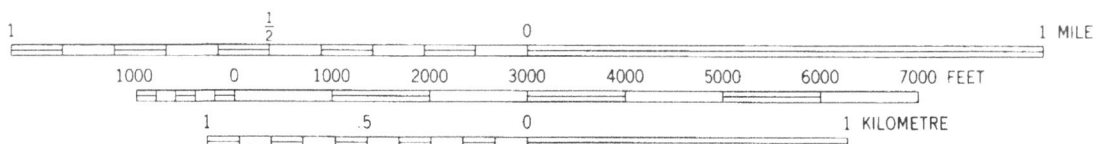

SCALE 1:24,000

Contour Interval 10 Feet

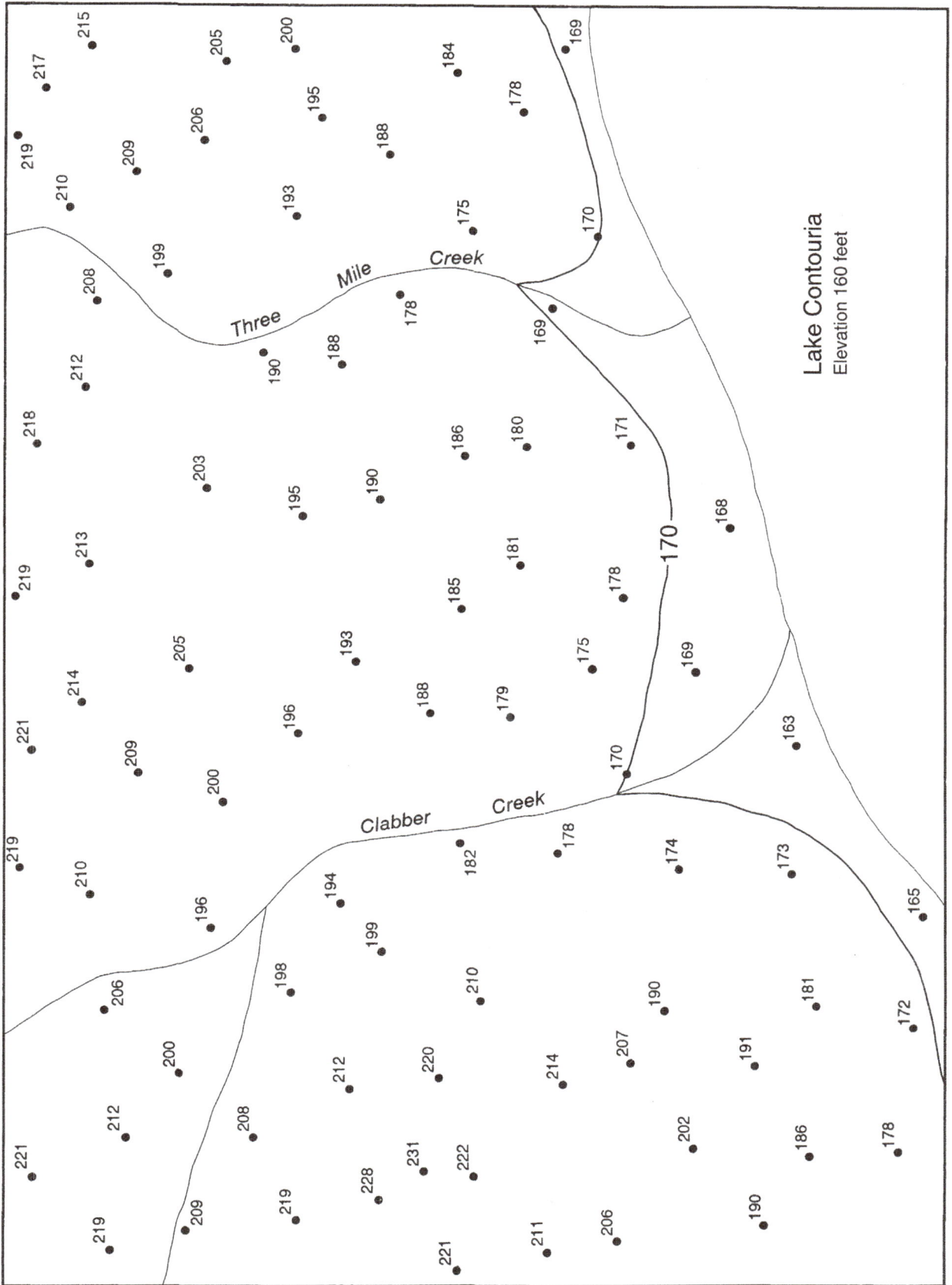

FIGURE 14.4 Map with elevation points.

Refer to the Stovepipe Wells map (Lab Exercise 24).

R.F.: _____

1. Determine the slope gradient between the mouth of Mosaic Canyon and the waste treatment plant south of Stovepipe Wells.

    rise (vertical distance) = _____ feet

    run (horizontal distance) = _____ feet

    a. Percent grade: _____

    b. Feet per mile: _____

    Stream Gradient can be calculated in the same way as percent grade, and is also expressed in feet per mile (or meters per kilometer). For example, you select a stream or stream segment that is 10 miles long (run). By using the contours at the lower and upper ends of the stream the vertical difference is 140 feet (rise). Therefore:

    $$\frac{140 \text{ ft elevational difference}}{10 \text{ mile distance}} = 14 \text{ ft per mile}$$

2. Refer to the Heidelberg, Kentucky map (Lab Exercise 21). Cave Branch is a small stream south of Mayse School that flows through rugged terrain before entering the Kentucky River. Determine the stream gradient of Cave Branch, taking into account bends in the stream and meanders when measuring stream length.

    Elevation difference (rise): _____ feet

    Stream length (run): _____ feet

    a. Percent grade: _____

    b. Feet per mile: _____

Refer to the Ripley South map. A dam is located on Hyde Creek, just south of Ripley, TN.

1. Approximately what is the water surface elevation of the reservoir?

    _____

2. Assume the dam is enlarged and heightened so that the water level reaches a height of 360 feet.

    a. How far from the dam will water backup into the valley?

    _____

    b. Approximately how many square miles will the surface of the reservoir cover?

    _____

c. Will any buildings be destroyed if the water rises to this elevation (360 feet)?

_____

How many? _____

d. What is the length of the roadway that will be flooded?

_____ feet

Aerial Photograph Interpretation

Aerial photographs are among the most useful tools for physical geographers. They can be used for mapping, city and regional planning, forest inventories, landscape evaluation, studies in earth sciences, and have many other purposes. Taking photographs of the same area (repeat photographs) over a long period of time can be used to determine both the rates and the extent of environmental change.

Interpretation requires determination of the photograph scale. The scale or representative fraction (R.F.) of a photograph (and of any map) indicates a relationship between the distance on the ground and corresponding distance on the photograph. For example, an R.F. of 1:12,000 means that one inch on the map or photograph equals 12,000 inches on the ground. Other distance measures (such as cm or feet) could also be used.

Refer to the _____ topographic map and corresponding aerial photograph in this exercise.

The topographic map has an R.F. of 1: _____

Based on this scale, you can determine the R.F. of the corresponding aerial photograph. Indicate the distance units you use (circle one): *inches* or *cm*

1. Select two landmarks or coordinates that can be easily located on both the topographic map and photograph.

2. Measure the distance between the two points on the photograph: _____
   (cm or inches). Enter this value under 4a.

   Determine the distance between these two points on the map: _____
   (cm or inches)

3. Convert the distance on the map to the real ground distance using the above R.F.

   Distance on the ground _____ (meters or feet). Enter this value under 4b.

4. The photograph scale is calculated by determining the ratio of the following values:

    a. Distance on the photo _____

    b. Distance on the ground _____

    R.F. = 4a : 4b

    Convert these values to equivalent units of measure.

5. Photograph R.F. (scale) is 1: _____

6. Determine the scale (R.F.) of the aerial photographs that correspond to the following topographic maps in this exercise:

    a. Topographic map _____ : map R.F. 1: _____

       air photo R.F.: 1: _____

    b. Topographic map _____ : map R.F. 1: _____

       air photo R.F.: 1: _____

    c. Topographic map _____ : map R.F. 1: _____

       air photo R.F.: 1: _____

    d. Topographic map _____ : map R.F. 1: _____

       air photo R.F.: 1: _____

7. Determine the distance on the ground between the designated coordinates on the aerial photographs that correspond to the following topographic maps in this exercise:

    a. Topographic map _____

       points: (1) _____ and (2) _____

       _____ feet

       _____ miles

       _____ meters

       _____ kilometers

    b. Topographic map _____

       points: (1) _____ and (2) _____

       _____ feet

       _____ miles

       _____ meters

       _____ kilometers

8. Determine the area on the ground that corresponds to the designated area on the aerial photographs and topographic map in this exercise:

   a. Topographic map _____

      Scale of the corresponding aerial photograph:

      1: _____

      Area on the photograph is _____ × _____ (in. or cm).

      Corresponding area on the ground:

      _____ feet^2

      _____ mile^2

      _____ m^2

      _____ km^2

   b. Topographic map _____

      Scale of the corresponding aerial photograph:

      1: _____

      Area on the photograph is _____ × _____ (in. or cm).

      Corresponding area on the ground:

      _____ feet^2

      _____ mile^2

      _____ m^2

      _____ km^2

Stereoscopic Vision

Aerial photographs can be viewed in stereoscopic (or three-dimensional) vision. The ability to visualize the relief of a site enhances your ability to study photographs. Aerial photographs are taken from low flying aircraft. Photographs are usually taken so that they overlap. They can be viewed in stereoscopic vision by using a stereoscope to look at two different photographs at the same time.

The two pictures are placed with one overlapping the other so that the feature you want to view is about the same distance apart as your eyes (2–2½ inches). A stereoscope is placed over the photographs and allows you to look at the same spot on the two photographs simultaneously. The angular difference between the camera in the airplane and the positions on the ground when the picture was taken creates slightly different photographs which allows three-dimensional viewing, but with significant vertical exaggeration.

Refer to Figure 14.5 which is a stereogram of a steep-sided volcanic cone. This stereogram was chosen because it can easily be seen by the first time stereo viewer.

Stereoscopes will be provided by the lab instructor. Place them over the stereogram. While looking through the stereoscope, slowly adjust its position until you see a clear three-dimensional image of the landscape.

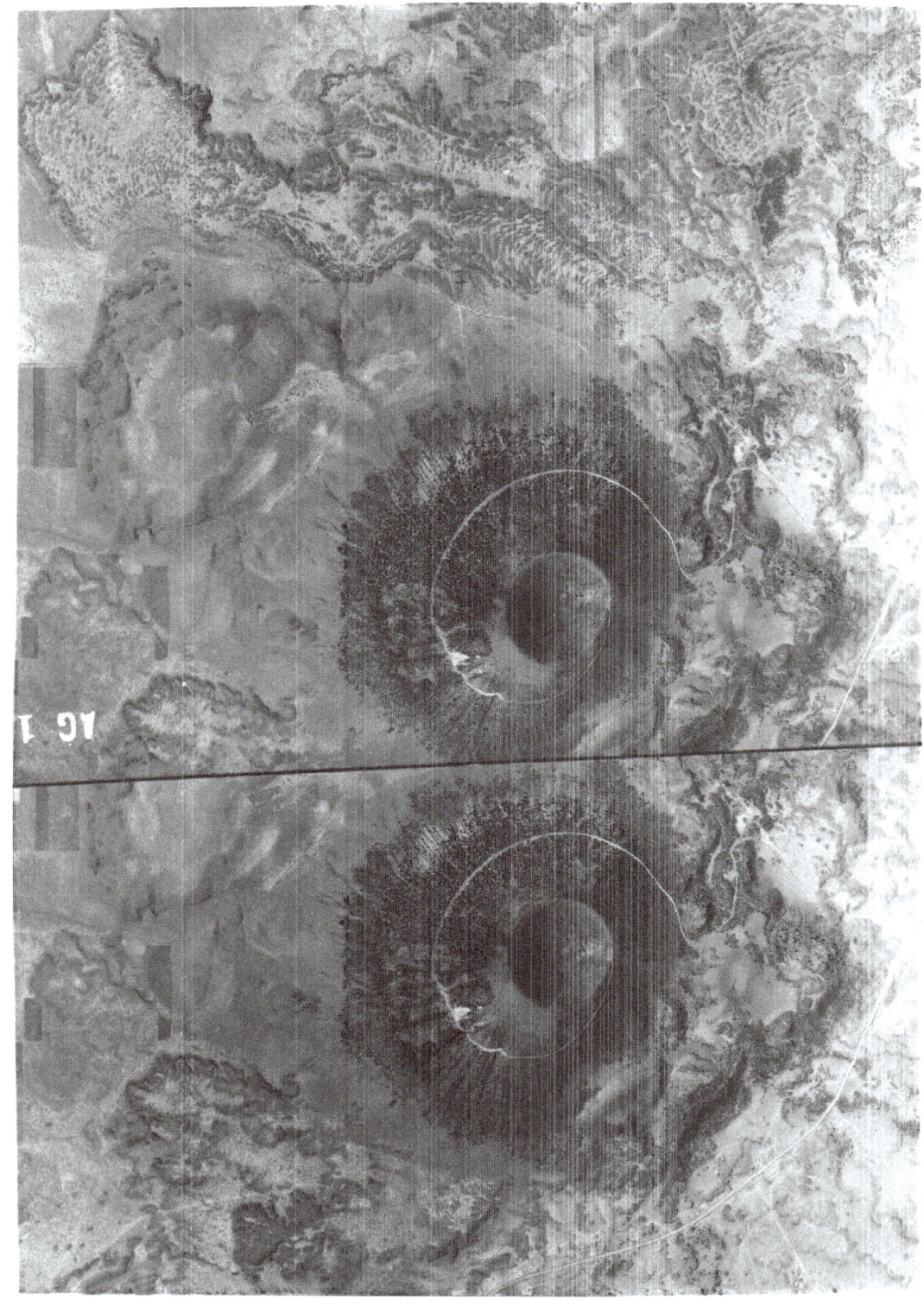

FIGURE 14.5 Mt. Capulin in New Mexico. Stereogram provided courtesy of the Map Library at the University of Illinois at Urbana-Champaign.

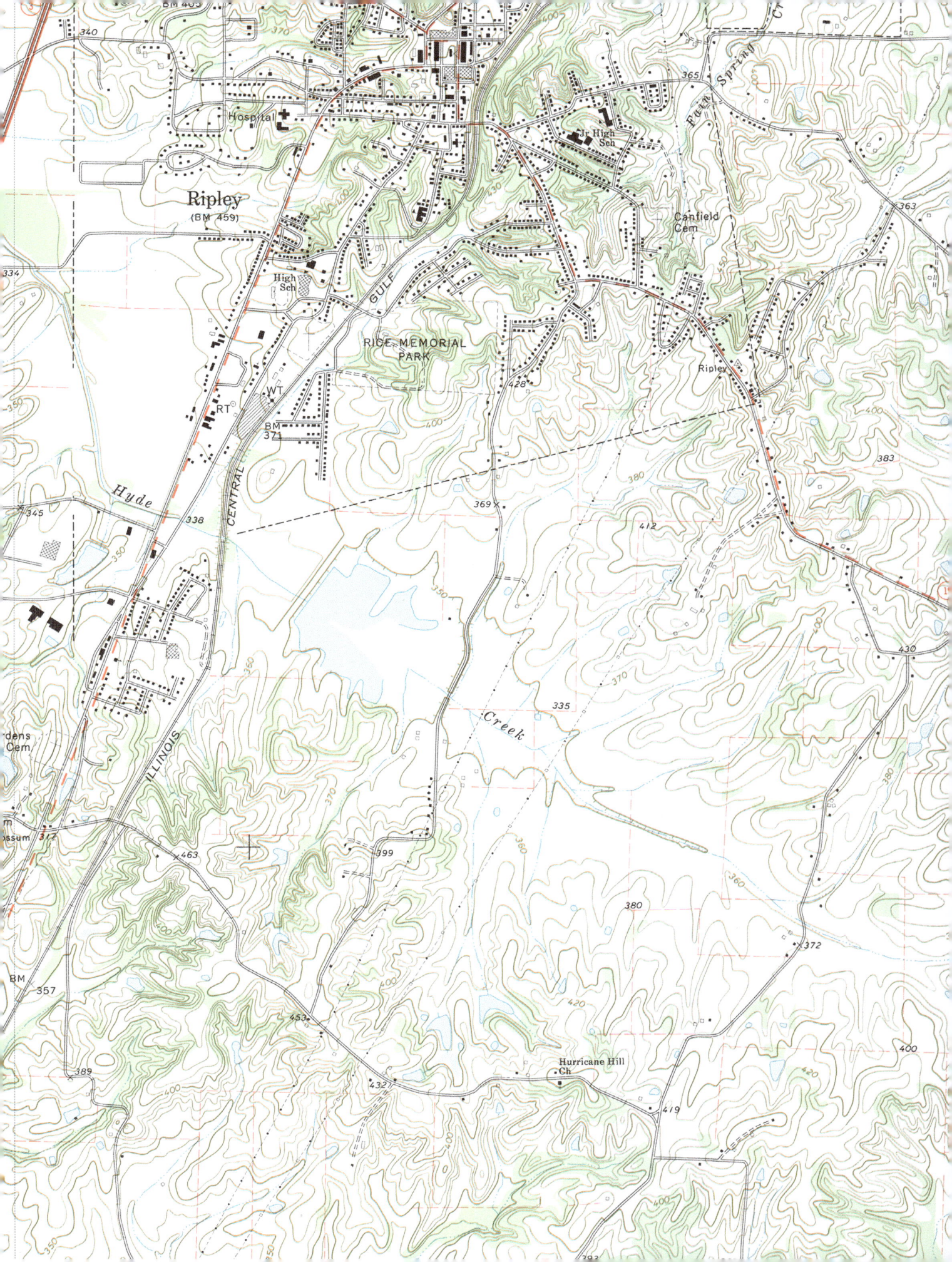

RIPLEY SOUTH, TENNESSEE

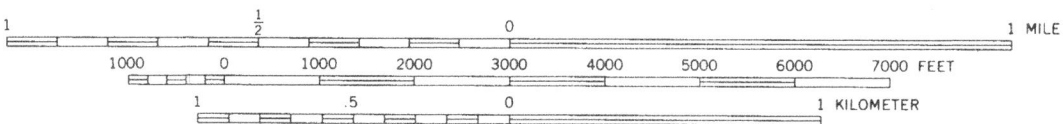

SCALE 1:24,000

Contour Interval 10 Feet

Name _____ Section _____

Global Positioning System Applications

LAB EXERCISE 15

Equipment

- GPS receivers

The global positioning system (GPS) is a satellite based system used to accurately determine positions at the earth's surface. It was developed by the U.S. Department of Defense for military applications. GPS is based on 24 high-altitude satellites located so that at least four can view any point on the earth's surface. The satellites transmit position information that can be picked up by receivers. The distance between the satellites and the receiver is used to determine position on the earth's surface and altitude. There are many types of GPS receivers. Most can be used to determine position to within 10m. A GPS receiver can indicate position using many different coordinate systems. Latitude and longitude, the most common coordinate system, is described in Exercise 2.

Universal Transverse Mercator Coordinate System (UTM)

The Universal Transverse Mercator (UTM) coordinate system is commonly used when determining a precise position with a GPS receiver. UTM uses north-south strips that divide the earth into pole-to-pole zones that are 6° longitude wide. The first zone begins at the 180° meridian (the international dateline) and runs east to the 174°W meridian. The other zones occur in contiguous increments of 6° longitude. So the second zone is from 174°W to 168°W, and so on. The number given to each zone increases with distance to the east. For the U.S., California on the west coast is within zones 10 and 11, while Maine on the east coast is in zone 19.

FIGURE 15.1 UTM zones in the U.S. contiguous states.

The UTM coordinate system indicates a position using meters. In the northern hemisphere this system uses distance in meters beginning from the equator called *northings*. In the southern hemisphere, northings measure distance from the South Pole. The circumference of the earth is about 40 million meters, so the distance from the equator to the North Pole, one-quarter circumference, is about 10 million meters. A position 5 million meters north is about one-half the distance from the equator to the North Pole.

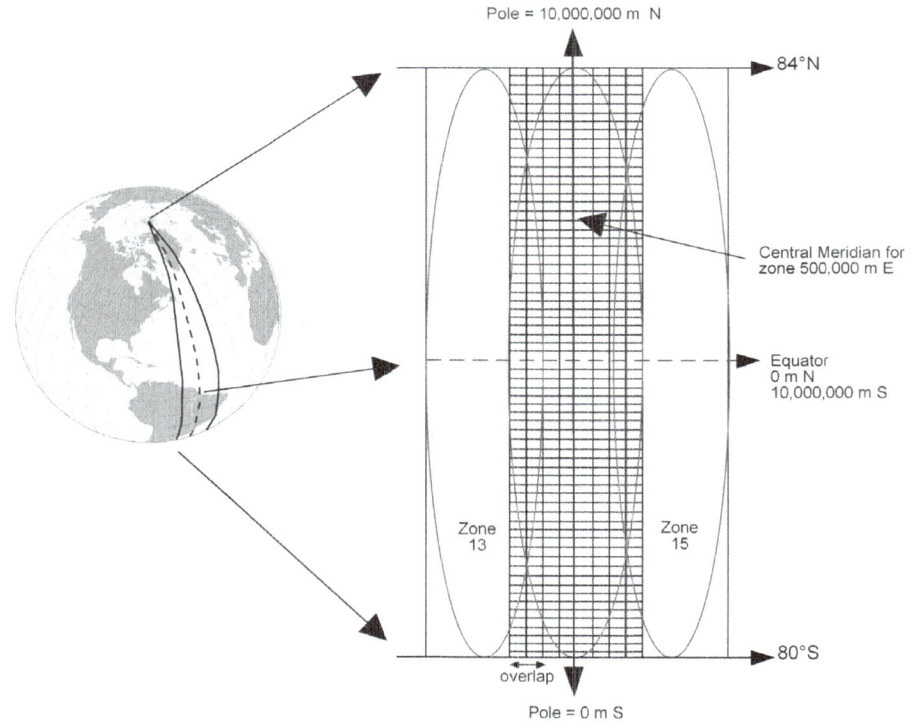

FIGURE 15.2 UTM coordinate system.

170

Eastings are based on the number of meters east of a *false origin* that is about half a degree beyond the westerly limit of each zone. The false origin was established so that the central meridian of a zone has an easting of 500,000m. An easting of less than 500,000m is west of the central meridian and an easting of greater than 500,000m is east of the central meridian.

Here is an example of UTM coordinates used to locate a position in zone 16. The state capital dome in Montgomery, Alabama is 3,581,300 meters north of the equator. It is also 566,000 meters east of the false origin used for zone 16, or 66,000 meters east of the central meridian that is 87° west of Greenwich (500,000 + 66,000 = 566,000). Note that GPS receivers usually show the zone number (in this case the zone is 16). The easting is reported first, followed by the northing:

16 566,000 3581300

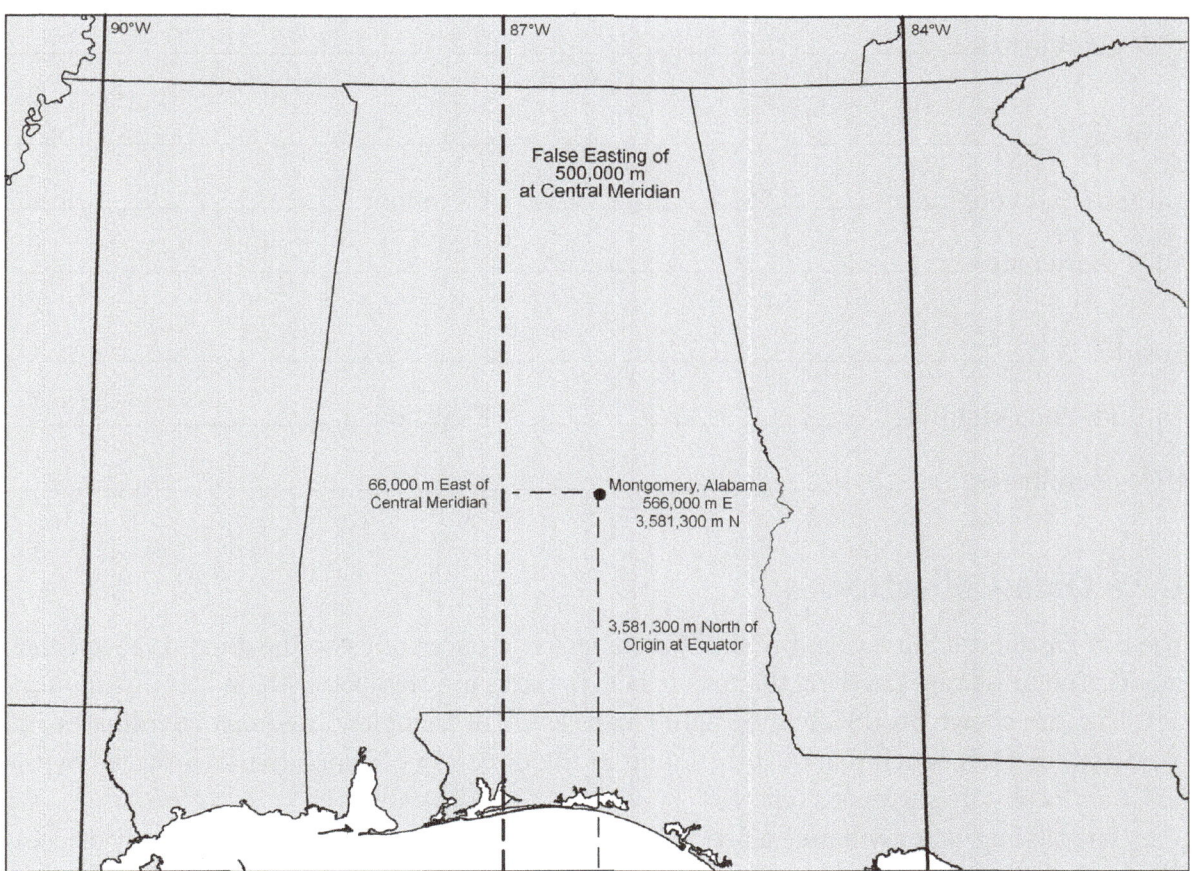

FIGURE 15.3 UTM coordinates of the capital dome in Montgomery, Alabama.

U.S. Geological Survey topographic maps show the UTM grid which is indicated by blue tick marks at the margin of the map. The tick marks are spaced every 1000m. UTM grid lines are typically not shown. A topographic map will be provided by the lab instructor.

The map: _____
(name of quadrangle)

Contour interval: _____

Position (or identifying features) on the map will be provided by the lab instructor. Use the map to determine coordinates and elevation for the following positions:

Position 1: _____

Latitude and Longitude: _____ Elevation: _____

UTM coordinates: _____

Position 2: _____

Latitude and Longitude: _____ Elevation: _____

UTM coordinates: _____

Position 3: _____

Latitude and Longitude: _____ Elevation: _____

UTM coordinates: _____

GPS Data Collection

The U.S. Geological Survey has installed permanent reference points on the ground called bench marks. The brass caps covering the tops of bench marks are stamped with an identifying number. They are shown on USGS topographic maps as small triangles. The exact coordinates and elevation for each bench mark can be found in the directory of bench marks provided by the USGS or local state surveyor's office.

The lab instructor will note a bench mark accessible from your location and provide coordinates and elevation.

Bench mark number: _____

	UTM coordinates	*Elevation (m)*
Bench mark information (from the USGS)	_____	_____

The coordinates and elevation shown on the GPS receiver may change slightly, even when the receiver is held stationary. Below, record four readings from the GPS receiver indicating UTM coordinates and elevation.

		UTM coordinates	*Elevation*
GPS receiver	1.	_____	_____
	2.	_____	_____
	3.	_____	_____
	4.	_____	_____
average:		_____	_____

Compare the exact location and elevation of the bench mark with average values from the GPS receiver. Note the difference in location and elevation in meters.

_____ (m)   _____

UTM coordinates will be given for the following positions. The lab instructor may indicate the general area in which you will look for the exact position. Use a GPS receiver to find and identify these positions (or identifying features).

1. UTM coordinates: _____

   General area: _____

   Position (identifying feature): _____

2. UTM coordinates: _____

   General area: _____

   Position (identifying feature): _____

3. UTM coordinates: _____

   General area: _____

   Position (identifying feature): _____

4. UTM coordinates: _____

   General area: _____

   Position (identifying feature): _____

The lab instructor will designate an area near your location. Use the GPS receiver at points on the perimeter of this area and note the general description and UTM coordinates of each below. Use the graph paper at the end of this exercise to map the designated area.

1. Perimeter location: _____

   UTM coordinates: _____

2. Perimeter location: _____

   UTM coordinates: _____

3. Perimeter location: _____

   UTM coordinates: _____

4. Perimeter location: _____

   UTM coordinates: _____

5. Perimeter location: _____

   UTM coordinates: _____

6. Perimeter location: _____

   UTM coordinates: _____

What is the length of the perimeter of the designated area? _____

What is the area within the perimeter? _____

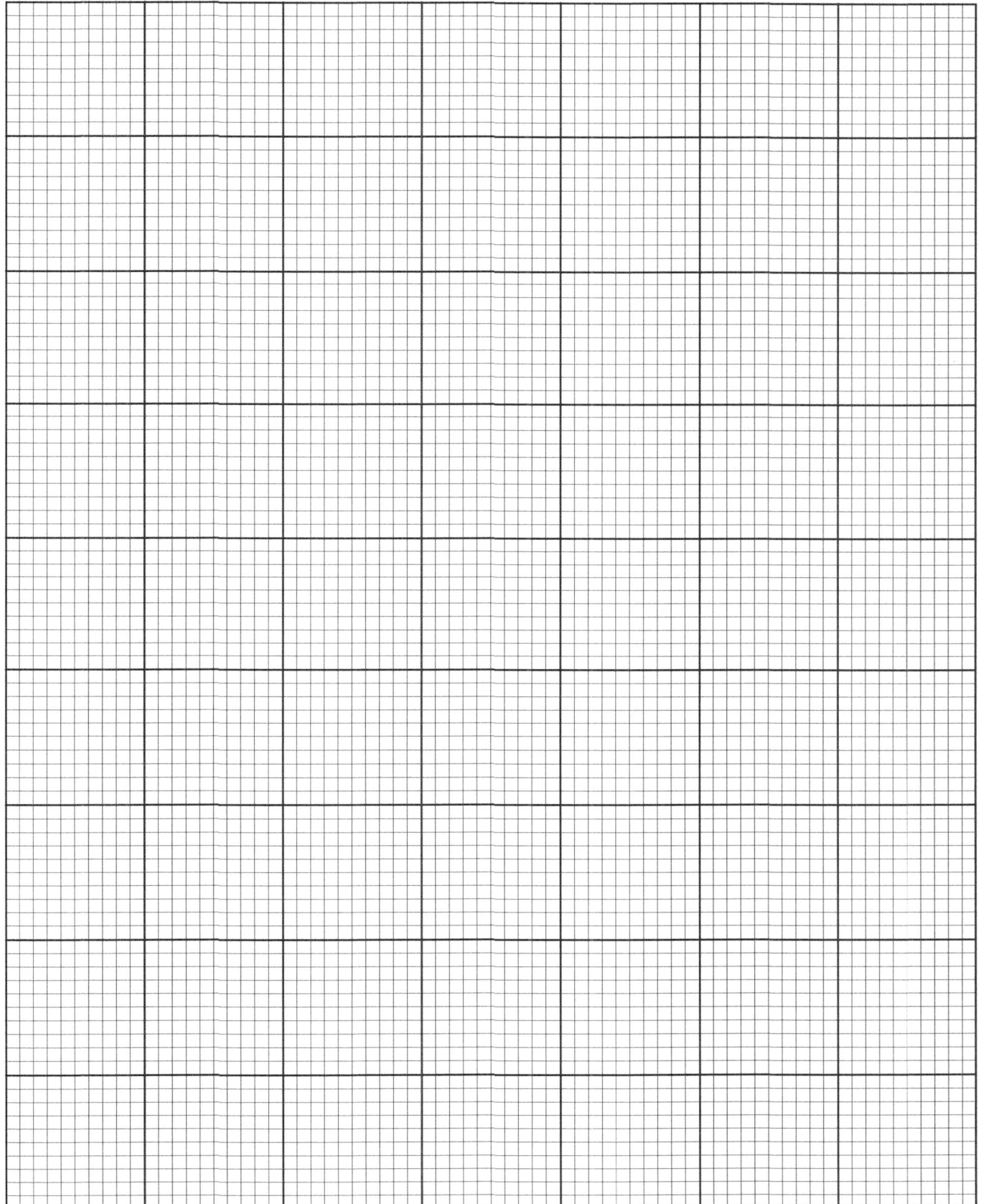

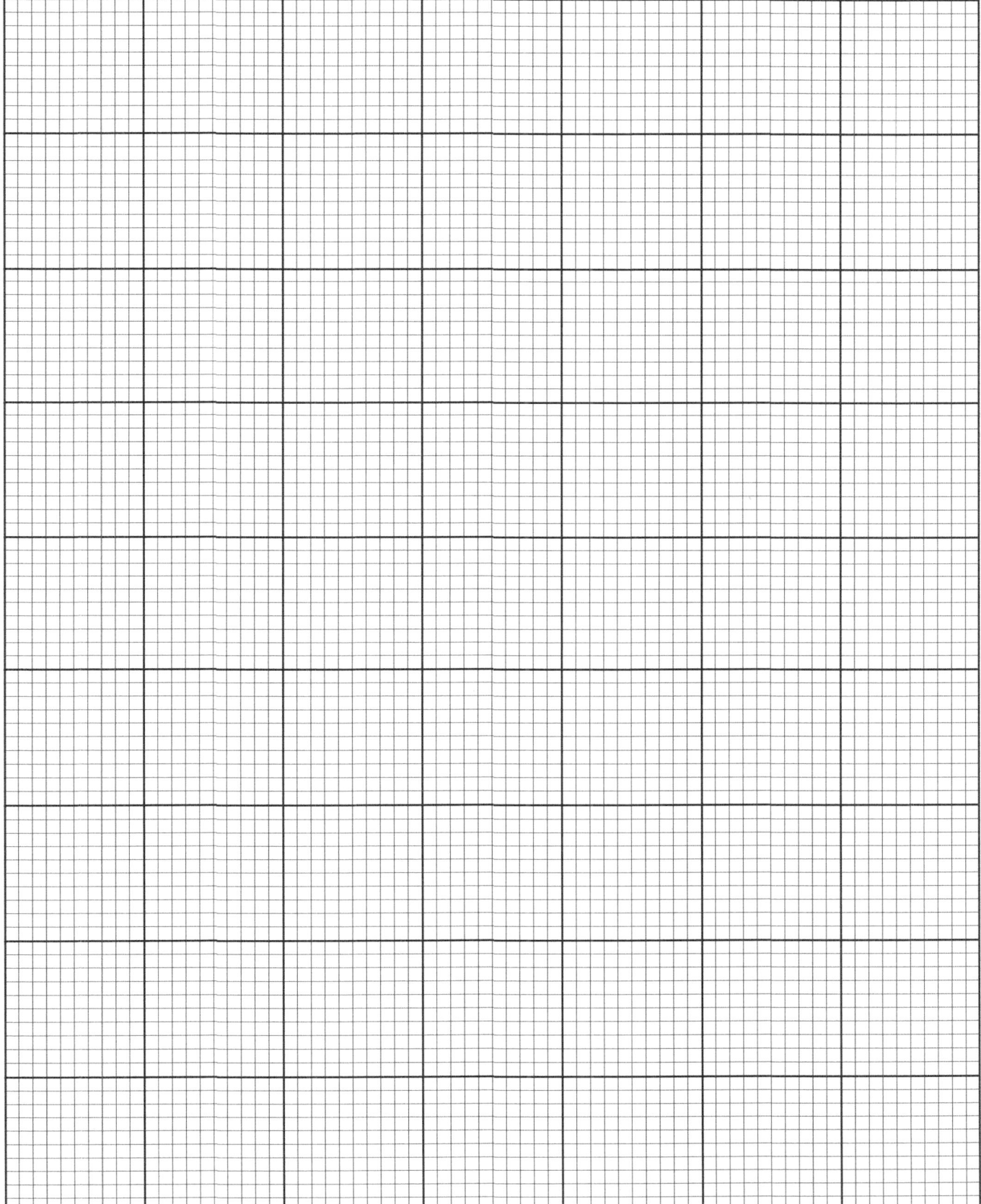

Name_____ Section_____

LAB EXERCISE 16

Trigonometry Applications in Geographical Field Work

Direct measurements of distance, area, or change in elevation, during fieldwork are often difficult or impractical. Trigonometry provides simple techniques to make indirect, but accurate calculations. In this exercise you will use (1) solutions for Right Triangles to determine both distance and height of objects and (2) solutions for Oblique Triangles to determine area.

Solutions for Right Triangles

Solutions for Right Triangles can be used to easily calculate heights of objects or changes in elevation. For example, you can use this technique to determine the height of trees, elevation of a cloud base, or elevation of a distant point relative to the elevation of your position. Refer to the Right Triangle shown below in Figure 16.1, where A, B, and C represent angles and a, b, and c are distances.

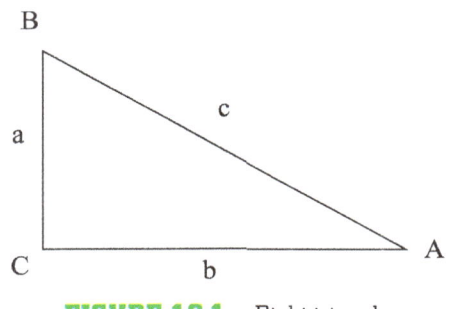

FIGURE 16.1 Right triangle.

TABLE 16.1 SOLUTION FOR RIGHT TRIANGLES

Given	Determine	Using the Following Solutions
A, a	b	$b = a * \cot A$
	c	$c = a / \sin A$
A, b	a	$a = b * \tan A$
	c	$c = b / \cos A$
A, c	a	$a = c * \sin A$
	b	$b = c * \cos A$

Note: Trigonometry tables are given in Appendix A.

179

1. Determine height of an object or change in elevation in Figure 16.2.

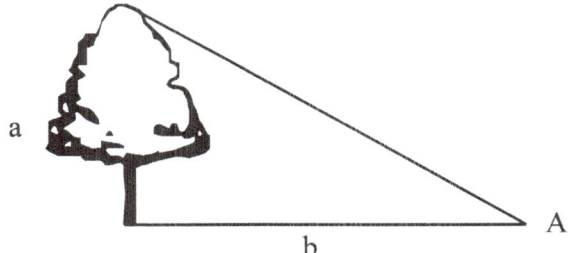

FIGURE 16.2   Determining height or change in elevation.

Assume you want to determine the height of a tree. Direct measurement usually is not possible. But you can use the solution for Right Triangles shown in Table 16.1. In this case, *a* is the height of the tree. You can easily measure your distance from the tree (*b*) and the angle from the ground to the treetop (A). Since you know A and *b*, use the solution shown above to determine tree height. In this case:

Height of the tree (*a*) = *b* * tangent of A

Distance to the tree (*b*)	Angle° (A)	Tangent of A	Height (*a*)
15m17	.3057		

You can use the same solution to determine the height of a distant point (building, hill or cloud base). If you know the horizontal distance to that point (*b*) and measure the angle (A), use the same solution shown above to determine the change in elevation, again referring to Figure 16.1.

Horizontal distance (*b*)	Angle° (A)	Tangent of A	Elevation
1km	11		

2. Determine distance using right triangles.

Using right triangles on maps or the ground, you can determine the distance between points. Assume you are standing on the bank of a river and want to know the distance to the opposite river bank as shown in Figure 16.3. You can use the same solution used to calculate height.

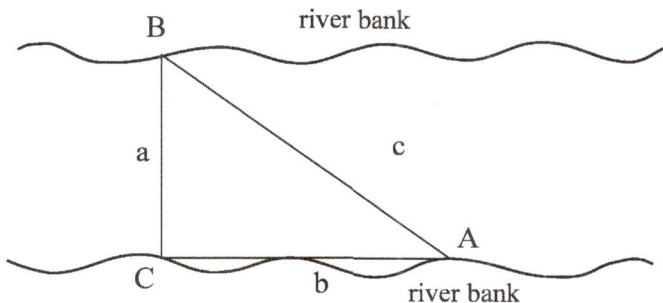

FIGURE 16.3 Using right triangles to determine distance.

In this case, a is the channel width. You can easily measure distance b on your side of the river and angle A. Use the following solution to determine distance across the river, as shown below:

Channel Width (a) = b * tangent of A

Distance along the bank (b)	Angle° (A)	Tangent of A	Channel width (a)
30m	47		

If you know the distance of any one side of the triangle, you can measure angle A and then use the solutions given in Table 16.1 to determine the distances of the other sides.

Given (a, b, or c)	Determine (a, b, or c)	Angle° (A)	Distance

Solution for the Area of Oblique Triangles

An oblique triangle is any triangle that is not a right triangle. It could be an acute triangle (all three angles less than 90°) or an obtuse triangle (one of the angles greater than 90°). Surface area on the ground or on maps can easily be determined by using oblique triangles. For example, you can calculate the area of any triangle on a map or on the ground by identifying the position of points A, B, and C, and measuring the distance between them. Refer to the oblique triangles shown on the following page in Figure 16.4.

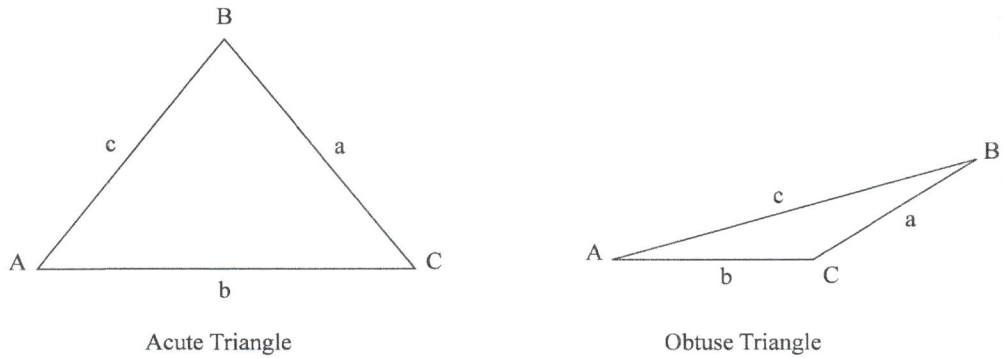

FIGURE 16.4 Oblique triangles.

Solution for the area of oblique triangles:

Given	Determine	Using the following solution	
a,b,c	area	area = $\sqrt{s(s-a)(s-b)(s-c)}$	where: s = (a + b + c)/2

Assume you want to determine a triangular area on the ground or corresponding area on a map. Measure the distance of each side of the triangle and then calculate the area using the above solution. **Note:** If the area is not triangular, you can use points on the perimeter that allow you to divide the area in contiguous triangles.

Distances			s	Area
a __15m__	b __12m__	c __11m__	_____	_____
a_____	b_____	c_____	_____	_____
a_____	b_____	c_____	_____	_____

Field Exercise

The lab instructor will indicate positions or objects for you to determine height (or elevation change), distance, or area using solutions for right and oblique triangles. Maps can be used if field observations are not possible.

Plot positions given by the lab instructor and triangles used to find solutions on the graph paper included at the end of this exercise. Draw the map with a pencil, not a pen, in case corrections are needed. The quality of your maps depend on accuracy! Use an engineers scale or ruler for all line work, and a protractor to determine angles.

1. Calculate distance using a right triangle solution. First determine distance (*b*) and then Angle (A).

Your position (point of observation)	Determine distance to following positions or objects
1. _____	_____
2. _____	_____
3. _____	_____

	Distance (b)	Angle° (A)	Tangent of A	Distance (a)
1.	_____	_____	_____	_____
2.	_____	_____	_____	_____
3.	_____	_____	_____	_____

2. Determine height (or elevation change) using right triangle solution. First, determine the horizontal distance to the position (b) and then angle A.

Your position (point of observation)	Determine distance to following positions or objects
1. _____	_____
2. _____	_____
3. _____	_____

	Horizontal Distance (b)	Angle° (A)	Tangent of A	Distance (a)
1.	_____	_____	_____	_____
2.	_____	_____	_____	_____
3.	_____	_____	_____	_____

3. Determine area using oblique triangle solution. First, you must identify and locate points on the perimeter of the designated area and create a triangle.

If the area is not triangular, use points on the perimeter that allow you to divide the area in contiguous triangles.

Determine the distance of each side of the triangle(s). Then find the area for each triangle and, if there is more than one triangle, add the values to determine the entire area.

Problem 1

Site description and boundary positions:

	Distances		s	Area
a_____	b_____	c_____	_____	_____
a_____	b_____	c_____	_____	_____
a_____	b_____	c_____	_____	_____
a_____	b_____	c_____	_____	_____
a_____	b_____	c_____	_____	_____

Total Area _____
(if combining the area of more than one triangle)

Problem 2

Site description and boundary positions:

	Distances		s	Area
a_____	b_____	c_____	_____	_____
a_____	b_____	c_____	_____	_____
a_____	b_____	c_____	_____	_____
a_____	b_____	c_____	_____	_____
a_____	b_____	c_____	_____	_____

Total Area _____
(if combining the area of more than one triangle)

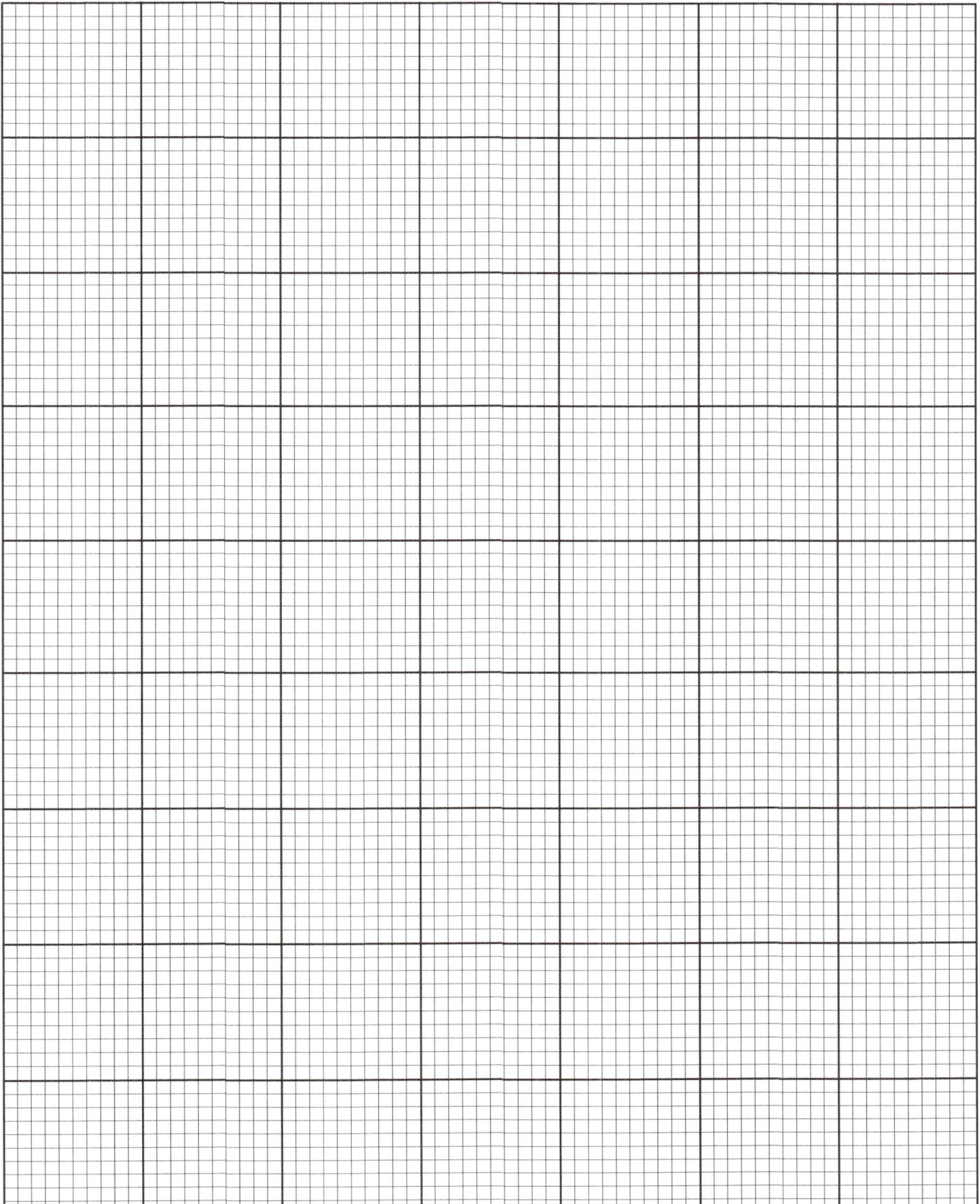

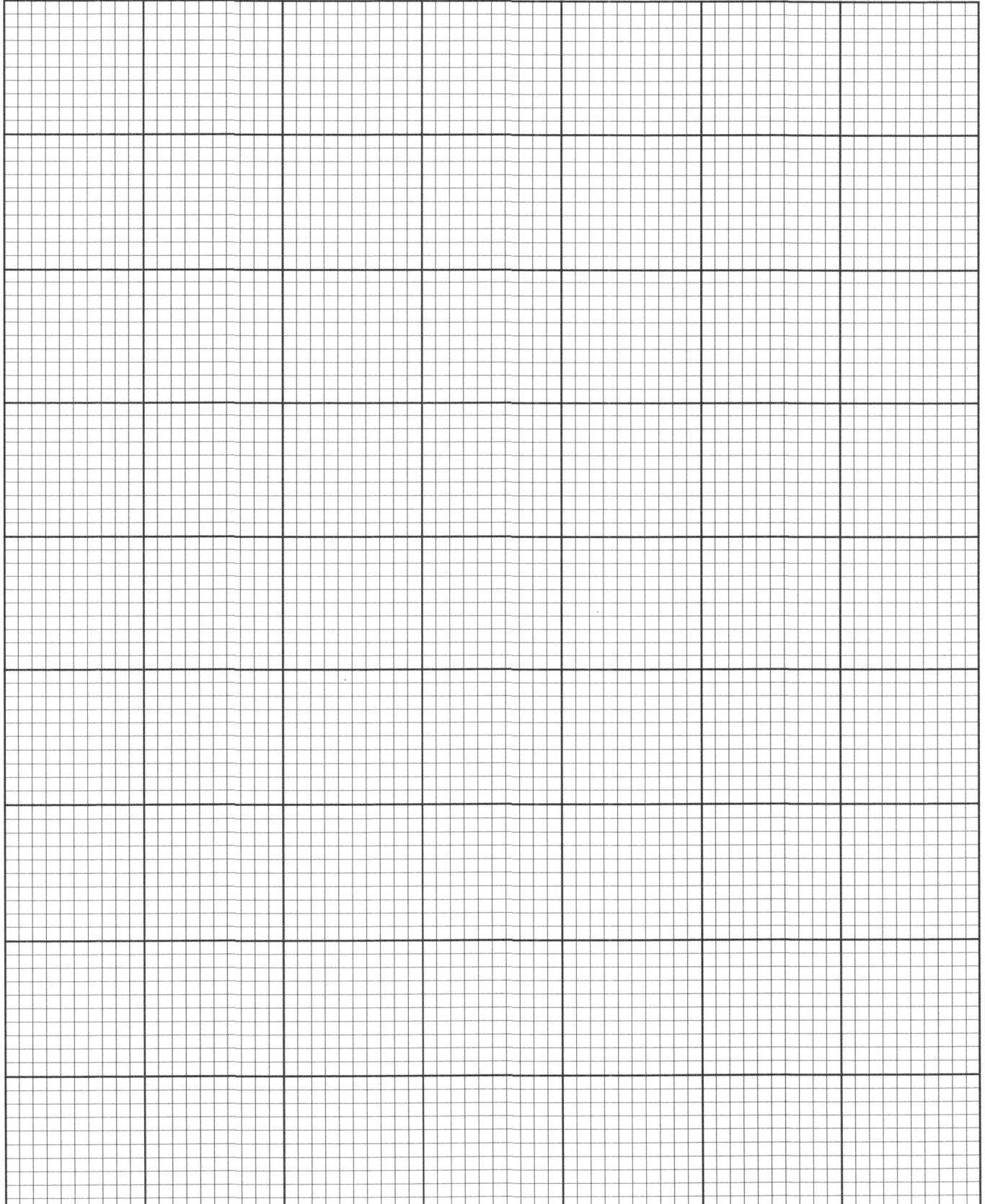

Name _____ Section _____

LAB EXERCISE 17

Contour Mapping

Equipment

- 3 brunton compasses with tripods
- 3 stadia rods
- 1 distance tape

Geographical field work occasionally requires the development of contour maps that show greater detail than the USGS topographic maps such as those used in Exercise 14. In this exercise you will collect field data from a small area designated by the lab instructor and use this information to create a contour map.

A sample data set is included at the end of this exercise. The lab instructor may use this data if an appropriate field site is not available, or to demonstrate how contour maps are constructed before you begin data collection.

Data Collection

1. The class will be divided into three groups, each working at a different compass.

   Establish 2 points (Station A and Station B) _____ feet apart along a designated line (baseline). Three compasses will be used. At Station A, 2 compasses will be set up: one for measuring inclination to determine the change in elevation or altitude relative to your position (group 1), and the second for azimuth or compass direction (group 2). At Station B the third compass will be used to determine azimuth (group 3).

   The study site will be within a rectangular area (designated by the lab instructor) adjacent to the baseline as shown in Figure 17.1.

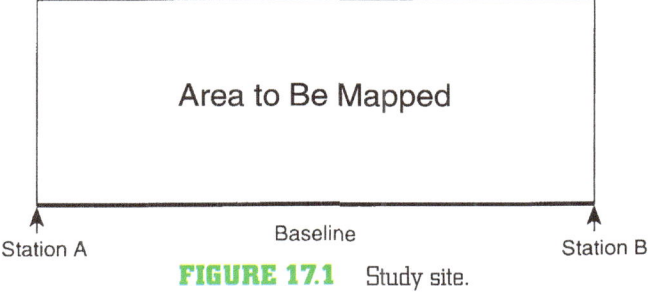

FIGURE 17.1 Study site.

189

2. One person from each group will take a stadia rod to a point within the study area. All three compasses will take readings based on this position. Azimuths will be determined from both Station A and Station B and entered on the following Table 17.1. The inclination (in degrees) will be determined from the third compass (at Station A) and also entered on Table 17.1. Everyone in each group will take a turn placing the stadia rod.

TABLE 17.1 DATA ENTRY

Station A is _____ feet elevation.

Names (observation points)	Station A (azimuth)	Station B (azimuth)	Inclination (Station A) (degrees)

Mapping

1. Create a map using either the data collected in the field (or you can use the data in Table 17.2 at the end of this exercise). Plot the map on the graph paper included at the end of this exercise.

   On the graph paper, all vertical lines will be oriented north-south and horizontal lines will be east-west. Draw the map with a pencil, not a pen, in case corrections are needed. The quality of your map depends on accuracy! Use an engineers scale or ruler for all line work, and a protractor to determine angles.

   The scale for this map is _____. Draw the baseline (a line from Station A to Station B). Remember that this line on the ground was _____ feet.

2. For each observation point, center the protractor on Station A and draw lines along the azimuths for the data. Repeat this using Station B. As shown in Figure 17.2, the intersection of each pair of lines determines the position of the point on the ground where the stadia rod was located.

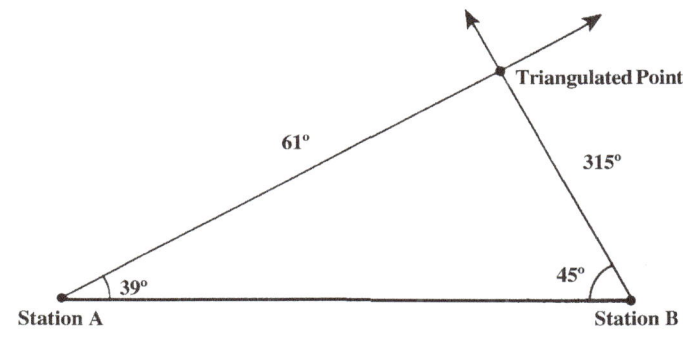

FIGURE 17.2 Mapping.

3. Measure the distance between all points and Station A and enter these values on the worksheet. These values should be calculated to the nearest 0.1 inch.

4. Use the map scale to determine the actual distance (X) on the ground. Enter the ground distances between all points and Station A on the worksheet.

5. Calculate the Y value (the difference in elevation from the compass that recorded inclination to a position 4 feet above the ground on the stadia rod). Determine the elevation of each point using the following:

Tangent of Angle a = Y/X
Therefore: Y = X * Tangent of Angle a

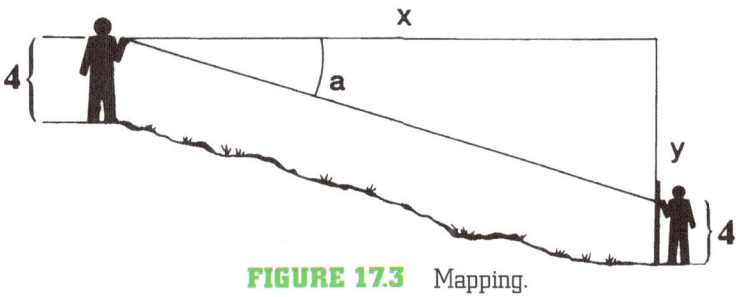

FIGURE 17.3 Mapping.

6. Construct a contour map by plotting the actual elevation for each survey point and replacing the point number with actual elevation. Draw contour lines using a contour interval of _____ ft., beginning with _____ ft. elevation.

The map must include a title, scale, date, direction orientation, and key (contour interval).

WORKSHEET

Point	Inclination	Measured Distance	Actual Distance(X)	Tangent	Y	Actual Elevation
1						
2						
3						
4						
5						
6						
7						
8						
9						
10						
11						
12						
13						
14						
15						
16						
17						
18						
19						
20						
21						
22						
23						
24						
25						

Sample Data Set

Table 17.2 is a sample data set that can be used to demonstrate how triangulation and inclination data can be used to construct a contour map. To use this data set, establish a north-south running baseline with Station A at the south-end of the baseline and Station B at the north-end. The study (mapped) area will be on the west side of the baseline.

Station A is _____ feet elevation

Refer to the mapping instructions on the preceding pages. A worksheet and graph paper for constructing the map are on the following pages.

TABLE 17.2 SAMPLE DATA SET

Observation Point	Azimuth A	Azimuth B	Inclination (Station A)
1	280	190	d9
2	300	200	d11
3	310	210	d11
4	320	230	d10
5	335	260	d12
6	0	180	d10
7	280	200	d11
8	280	210	d10
9	300	190	d10
10	300	210	d11
11	300	230	d8
12	310	190	d9
13	310	200	d12
14	310	230	d9
15	320	190	d10
16	320	200	d12
17	320	210	d11
18	335	190	d10
19	335	200	d11
20	335	210	d10
21	335	230	d11

WORKSHEET FOR SAMPLE DATA

Point	Inclination	Measured Distance	Actual Distance(X)	Tangent	Y	Actual Elevation
1						
2						
3						
4						
5						
6						
7						
8						
9						
10						
11						
12						
13						
14						
15						
16						
17						
18						
19						
20						
21						
22						
23						
24						
25						

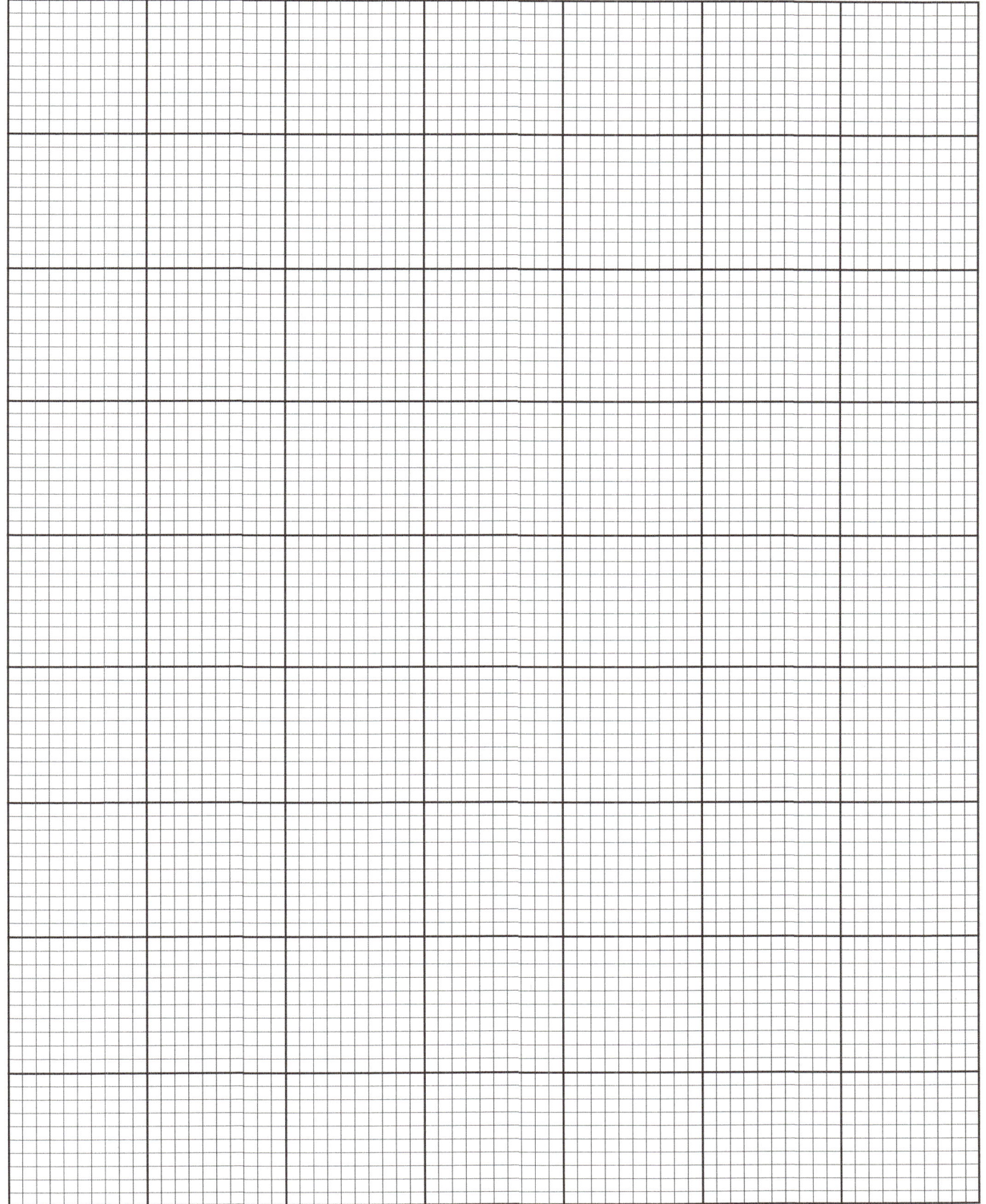

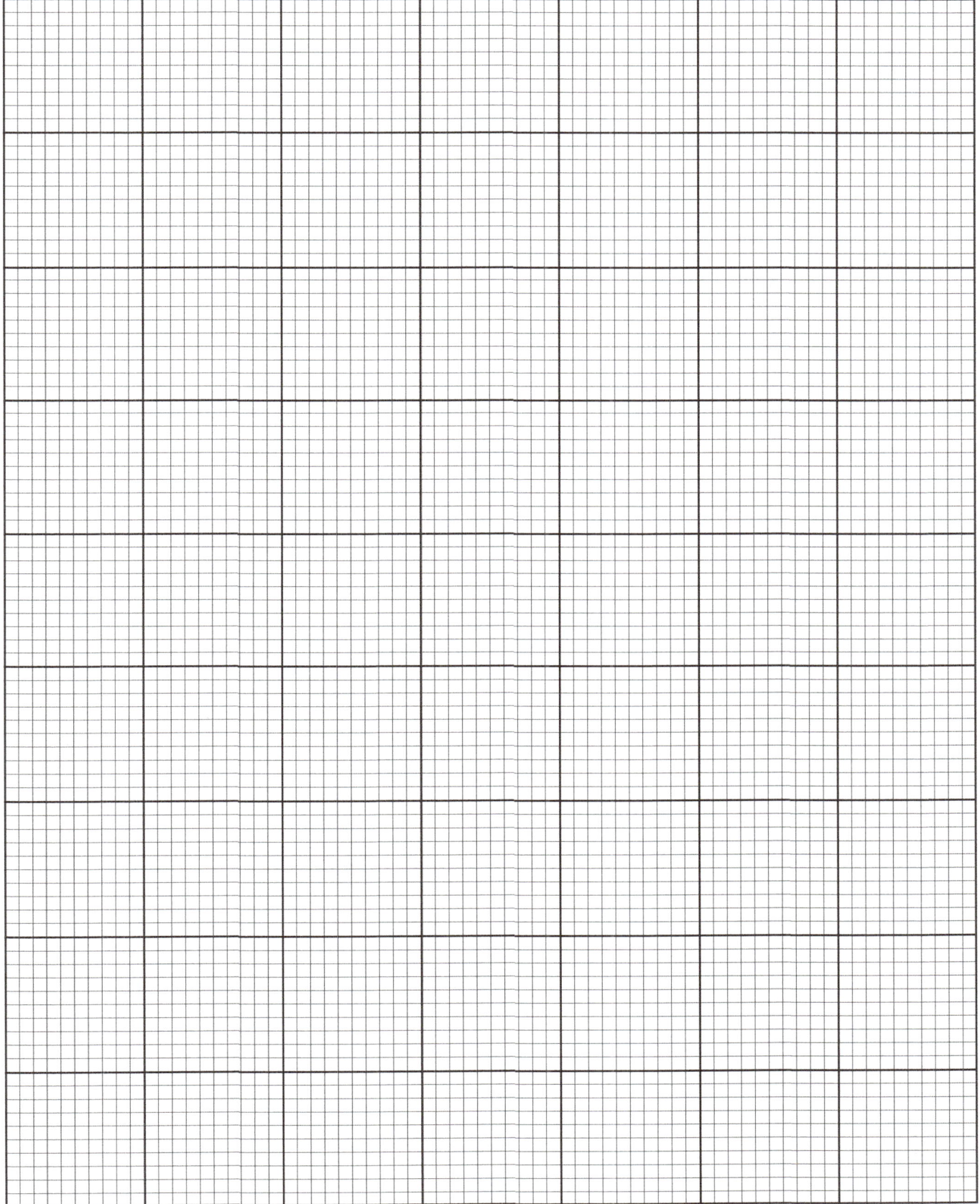

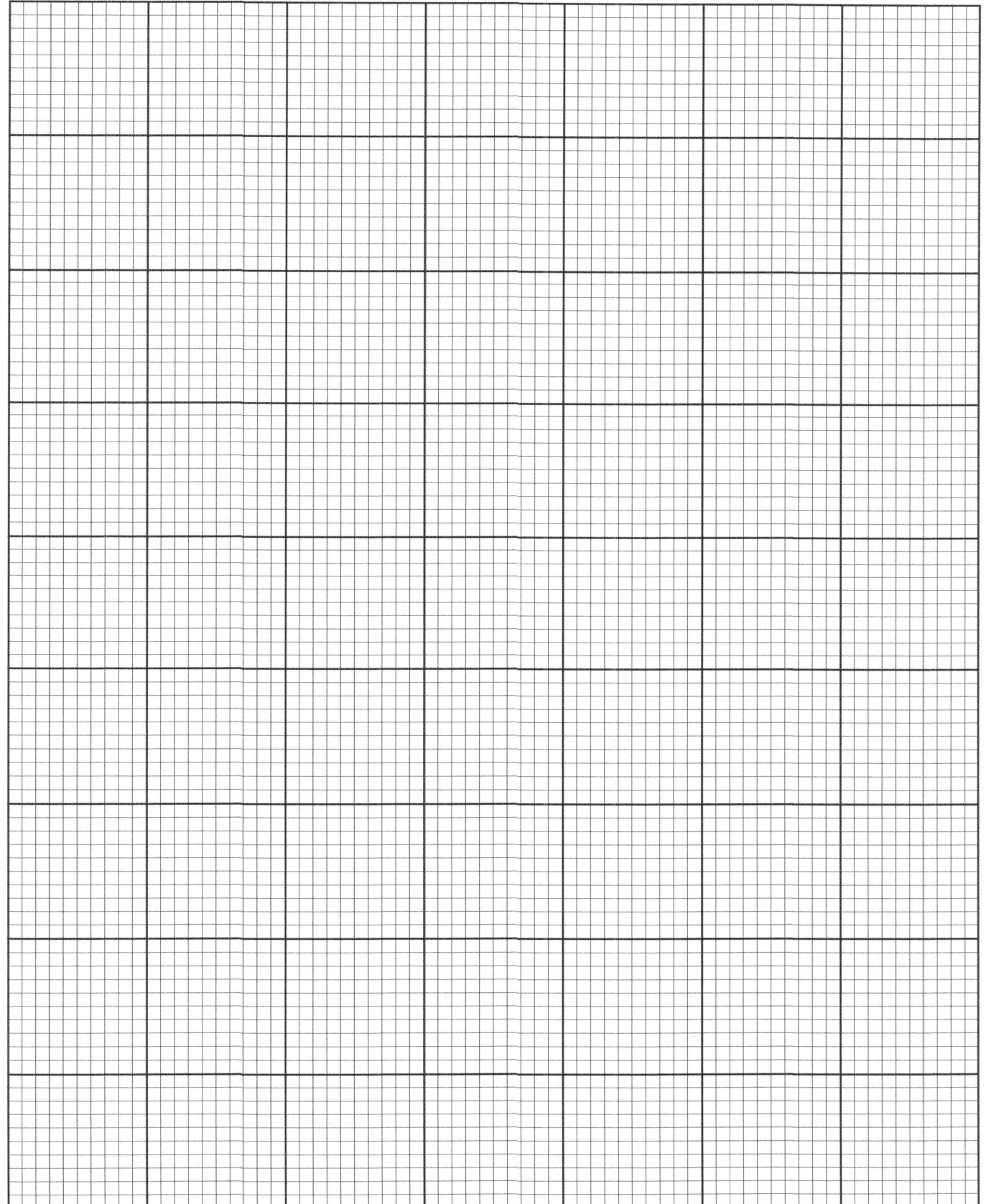

Name _____   Section _____

LAB EXERCISE

18

Soil Properties

Soil refers to the unconsolidated material occupying the earth's surface which provides a medium for plant growth. Physical and biological activity in the upper levels of soils results in characteristics that distinguish soils in one region from another. Five factors are considered essential in the development of soils: parent material, climate, vegetation, topography, and time. Parent materials are important to the physical and chemical makeup of young soils, but its influence is gradually reduced with increasing maturity.

Climate is the most important developmental factor in mature soils. Climate affects the rate of weathering, amount of chemical activity, the abundance of microorganisms, in addition to the vegetative cover and amount of organic material at the soil surface. Topography is also important in soil development. Slope and drainage can result in distinct soil characteristics even though parent material, climate, and vegetation are similar. Time is included among factors in soil development because of the direct relationship between the age of the development of soil profiles.

Soil Profiles

Soil profiles are sequences of horizontal soil layers called **soil horizons**. Horizons differ from those immediately above or below in physical, chemical, and biological properties. The major horizons are:

O: Organic horizons formed above mineral soil containing: (O_1) recognized forms of vegetation and (O_2) highly decomposed organic matter.

A: Mineral horizons consisting of: (A_1) a mixture of organic matter near the surface, (A_2) horizons that have been leached of clay, iron, and aluminum to lower horizons with a resulting concentration of sand and silt minerals, and (A_3) a lower zone similar to A_2, but transitional to lower horizons. The A horizon is a zone of **eluviation**, where soil material is moved downward to the lower horizons.

B: Horizons containing weathered material and an accumulation of iron and aluminum: (B_1) a zone transitional from A to B but containing material leached from above, (B_2) zone of maximum accumulation (**illuviation**), and (B_3) a lower zone similar to B_2 but transitional to C.

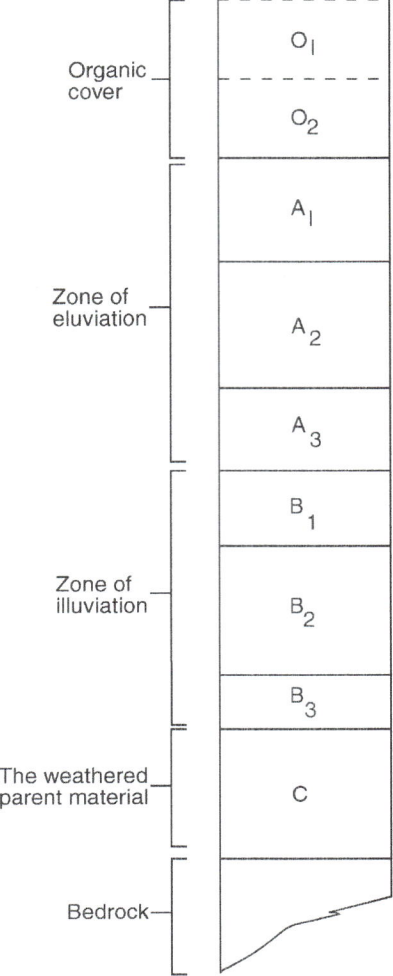

FIGURE 18.1 Idealized soil profile.

C: Zone of unconsolidated material, least weathering, and little organic material.

Soil profiles are highly variable depending on the five developmental factors listed above. There is a strong regional relationship between climate and vegetation, and both have a strong, direct effect on soil processes. Therefore, humid regions with forest vegetation typically have soils with similar properties, as do semi-arid grasslands, and deserts.

Soil Profile Descriptions

The lab instructor will determine an outdoor location to observe a soil profile. Record your visual observations on the data sheets at the end of this exercise. In addition, draw horizon boundaries, designations, and features on the graphs at the end of this exercise. On the drawings, include the depth (cm) of each horizon.

Afterwards, read the description of these soils from the County Soil Survey or information provided by the lab instructor and compare this information to your observations.

Physical Properties of Soils

The most important physical properties of soils are texture and structure. Texture is the proportion of different sized soil particles (sand, silt, and clay) and structure refers to the way in which the soil particles are grouped. Texture and structure determine the pore space in soils that affects rate of infiltration and water holding capacity, susceptibility to erosion, and ease of cultivation. Also, texture strongly affects chemical and biological processes. The size of soil particles are classified into three major groups: sand, silt, or clay. As shown below in Table 18.1, some subdivisions are also used.

TABLE 18.1 SOIL SIZE CATEGORIES (DEPARTMENT OF AGRICULTURE)

Classification	Diameter (mm)
gravel	Above 2.00
very coarse sand	2.00–1.00
coarse sand	1.00–0.50
medium sand	0.50–0.25
fine sand	0.25–0.10
very fine sand	0.10–0.05
silt	0.05–0.002
clay	below 0.002

Sand. Sand grains are large enough to be seen and can be felt when rubbed between the fingers. Because of the large grain size and uneven surfaces, sand particles make only limited contact with other surfaces. Therefore, sand is not sticky nor plastic and does not make stable aggregates in soils. The pore spaces in sand are very large and infiltration capacity and permeability to water and air is high.

Silt. Silt particles cannot be seen nor felt when rubbed between the fingers. Silt is more cohesive and adhesive than sand but has only limited stickiness. Silt has much smaller pore spaces than sand, and therefore, has lower permeability than sand. Like sand, however, it contributes little to aggregate formation in soils.

Clay. Clay particles are so small they can only be seen using an electron microscope. Extensive contact between clay particles compared to sand and silt results in stickiness and plasticity. Also, clay forms hard clods when dry. Pore spaces in clay are very small and therefore infiltration capacity and permeability is extremely low.

Soils are grouped into 12 textural classes that are shown in Figure 18.2. The textural classes identify the particle sizes that determine the properties of a soil. Loams are soils that have intermediate properties.

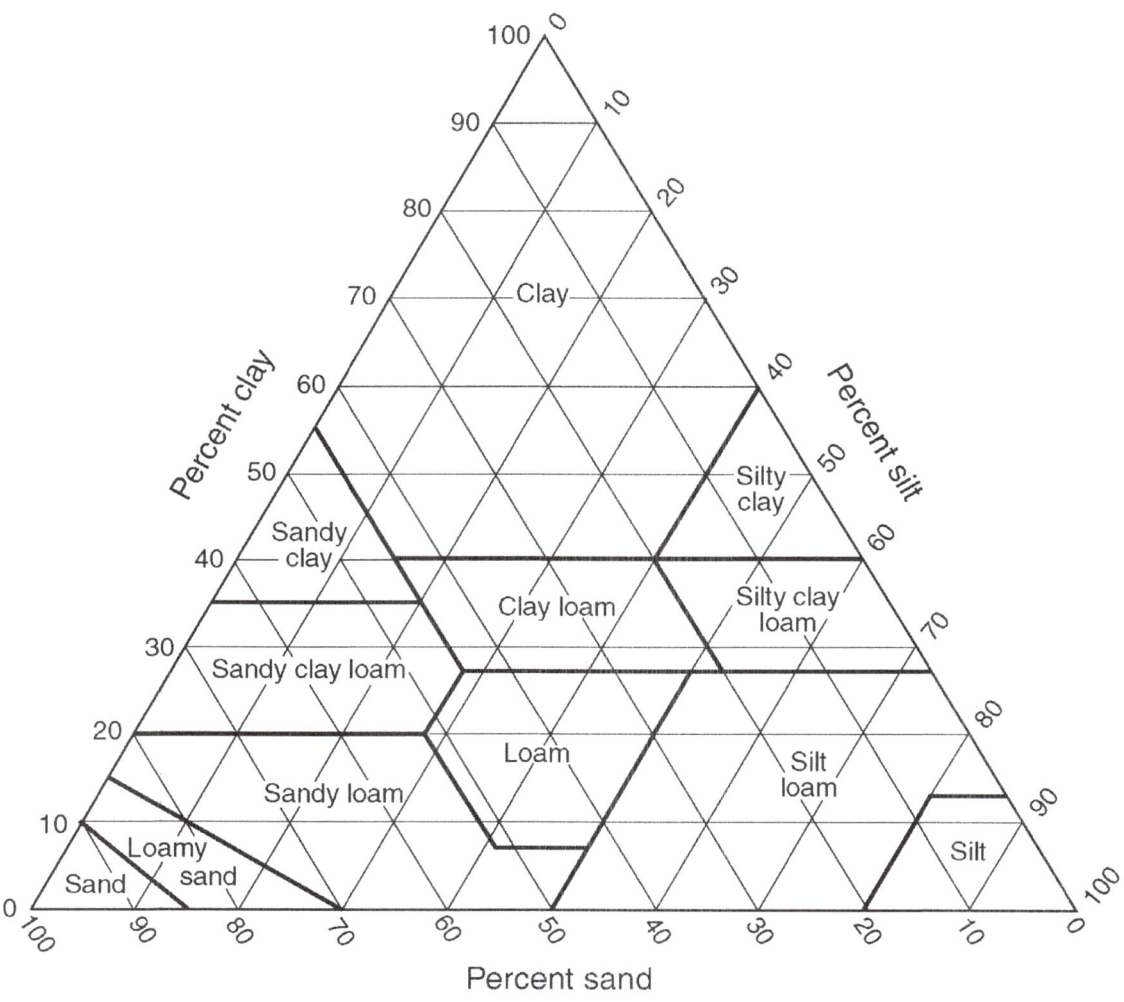

FIGURE 18.2 Soil textural classes. (U.S. Department of Agriculture.)

Determine Soil Texture

1. A rough approximation of soil texture can be made by using a graduated tube and timer. This method uses the same principle as standard scientifically more accurate methods: the rate of settling of soil particles in water. Soil samples will be collected or provided by the lab instructor.

   a. Fill about one-third of graduated cylinders with each of the soil samples.
   b. Shake or tap the soil down so that there are no large air spaces and record the number of cubic centimeters:

	Volume	Weight
sample 1:	_____ cc	_____ g
sample 2:	_____ cc	_____ g
sample 3:	_____ cc	_____ g
sample 4:	_____ cc	_____ g

c. In each of the cylinders add about 2 drops of Soil Dispersing Reagent for every 3 cubic centimeters of soil. Then, fill the cylinders with water to within about 5–6 cm of the top. Close or cap the top and shake vigorously for 2 minutes. Make sure the soil and water are thoroughly mixed.

d. Place the cylinder in a rack for exactly 30 seconds. *Be sure not to disturb the cylinder.* Then, measure the depth of the deposited material and record these values on the table below.

Let the solution stand for 30 minutes, then measure the depth of the deposited material again, and record these values on the same table. The first reading is the depth of the sand and the second is the depth of the sand and silt combined. The remaining particles are clay. The amount of clay can be determined by subtracting the total of sand and silt from total volume (cc).

	First Reading (cc) (sand)	Second Reading (cc) (silt and sand)	Remaining Particles (cc)(clay)
sample 1	_____	_____	_____
sample 2	_____	_____	_____
sample 3	_____	_____	_____
sample 4	_____	_____	_____

e. Determine the percentage of sand, silt, and clay for each sample by comparing the amount of each with the amount of the original soil. Refer to the Guide for the Soil Textural Classification (Figure 18.2) to determine the textural type of each sample.

	Percentage			Textural Classification
	sand	silt	clay	
sample 1	_____	_____	_____	_____
sample 2	_____	_____	_____	_____
sample 3	_____	_____	_____	_____
sample 4	_____	_____	_____	_____

f. Which of the samples has the largest pore spaces? _____

g. Which of the samples has: highest infiltration capacity? _____

highest permeability? _____

lowest infiltration capacity? _____

lowest permeability? _____

Explain:

h. Which of the samples has: highest water holding capacity? _____

lowest water holding capacity? _____

Explain:

2. Soil texture can be estimated by feel. Sand is coarse and gritty, silt has a smooth feel, and wet clay is sticky and somewhat plastic. The smallest particles you can see are coarse silt. Based on feel, estimate the textural class of soil samples provided by the lab instructor.

First, feel the soil when it is dry. Hard soil samples, those that do not easily crumble, contain at least a moderate amount of clay. Take a small soil sample and add water so that it has the consistency of modeling clay. Try to form a ribbon or wire by rolling the soil between your fingers. If a long thin wire can be easily formed, the soil is plastic and probably contains at least 40 percent clay. If a long ribbon can be formed but easily breaks, the soil is probably a clay loam or silty clay loam.

If the wet soil has a gritty feel, the soil has a sandy texture. If the feel is mostly like smooth flour and not very gritty, the soil has a silty texture. Wet sand will leave your hands moist but clean, a loamy sand (soils with a higher silt content) will leave them slightly soiled.

Textural Classification

sample 1 _____

sample 2 _____

sample 3 _____

sample 4 _____

Determine Bulk Density

Bulk density is the weight of a dry soil sample divided by its volume, or:

$$\text{Bulk Density (BD)} = \frac{\text{dry soil weight}}{\text{soil volume}}$$

and is usually expressed in grams/cm^3. The greater the bulk density, the less the pore space. For example, solid rock has no pore spaces, and therefore, a high weight per volume and high bulk density. In contrast, rock fragments will contain numerous pore spaces because they do not fit perfectly together. Therefore, the weight per volume and bulk density are comparatively less. Generally, bulk density is lowest near the soil surface and decreases with depth.

The bulk density of fine textured soils generally range from 1.0 to 1.3 g/cm^3 and for sandy soils from about 1.4 to 1.7 g/cm^3. Tillage loosens soils and lowers bulk density. Compaction will raise bulk density.

Undisturbed soil samples will be collected from the field or provided by the lab instructor. *These samples should not be compacted nor broken apart.* Metal ring samplers are used by soil scientists to remove undisturbed soil cores. If these are not available, you can use a shallow tin can with both ends removed, a block of wood, mallet, and soil sample bag.

a. Use a shovel or knife to smooth the area to be sampled.

b. Place the tin can on the soil surface and put the block of wood on top. Use the mallet to drive the can squarely into the soil, being sure to avoid compaction. Dig out the buried can and trim the excess soil from the bottom, again being careful to avoid compaction.

c. Remove the samples from the cans or sample bags and allow them to dry thoroughly.

d. Record the volume (cubic centimeters) and weight (after drying) of each sample on the table below. Calculate the bulk density by dividing the weight of the sample by its volume.

	Volume (cc)	*Weight (g)*	*Bulk Density (weight/volume)*
sample 1	_____	_____	_____
sample 2	_____	_____	_____
sample 3	_____	_____	_____
sample 4	_____	_____	_____

Particle Size and Surface Area

Chemical activity occurs on soil surfaces and therefore the amount of chemical activity is proportional to the amount of the surface area. Finer soils (those consisting of smaller particle sizes) have a greater total surface area. For example, a solid rock has a low surface:area ratio. If you crush the rock into many pieces, the total surface area for all the pieces is much higher, even though the volume has not changed. Therefore, the surface:area ratio is much higher. The same is true of sand that weathers into finer particles (silt or clay); the surface:area ratio of the soil increases.

1. How much surface area is exposed on a cube:

    1-cm on a side: _____ cm^2

2. What would be the total surface area if the 1-cm cube were divided into cubes:

	cm^2	m^2	particle size class
.01 cm on a side?	_____	_____	_____
.001 cm on a side?	_____	_____	_____
.0001 cm on a side?	_____	_____	_____

    Refer to Table 18.1 to determine the particle size class.

3. Which particle size class has the greatest chemical activity?

    _____

Name _____

Section _____

Soil Profile Data Sheet 1

Location: _____

Climate: _____
Vegetation: _____
Parent Material: _____
Topography: _____
Time (Developmental Stage): _____

Discernible horizon properties:

O: _____
_____
_____

A: _____
_____
_____

B: _____
_____
_____

C: _____
_____
_____

Soil Survey Report:

Series: _____
Order: _____
Capability Class: _____

Remarks:

Name _____

Section _____

Soil Profile Data Sheet 2

Location: _____

Climate: _____
Vegetation: _____
Parent Material: _____
Topography: _____
Time (Developmental Stage): _____

Discernible horizon properties:

O: _____
_____
_____

A: _____
_____
_____

B: _____
_____
_____

C: _____
_____
_____

Soil Survey Report:

Series: _____
Order: _____
Capability Class: _____

Remarks:

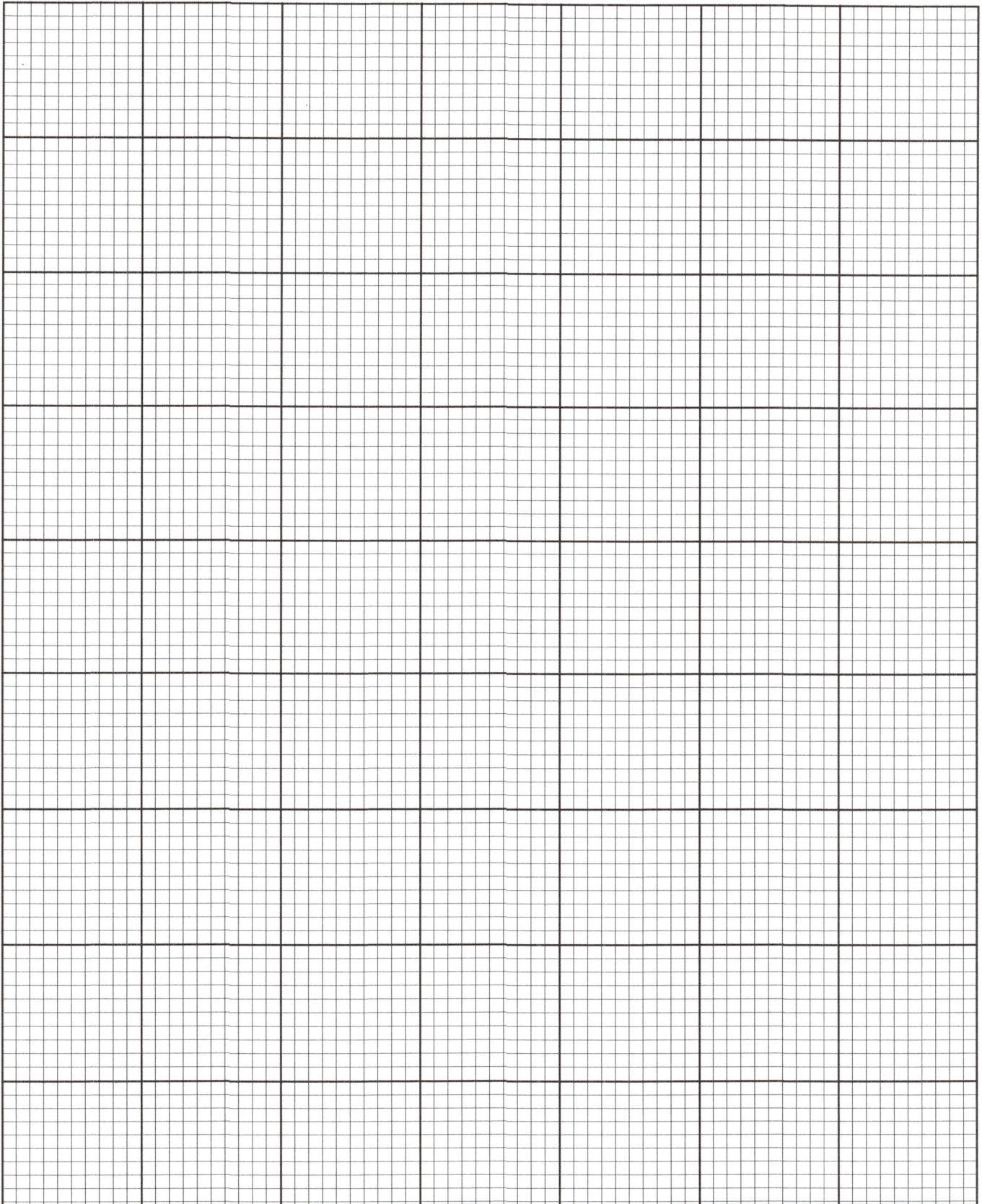

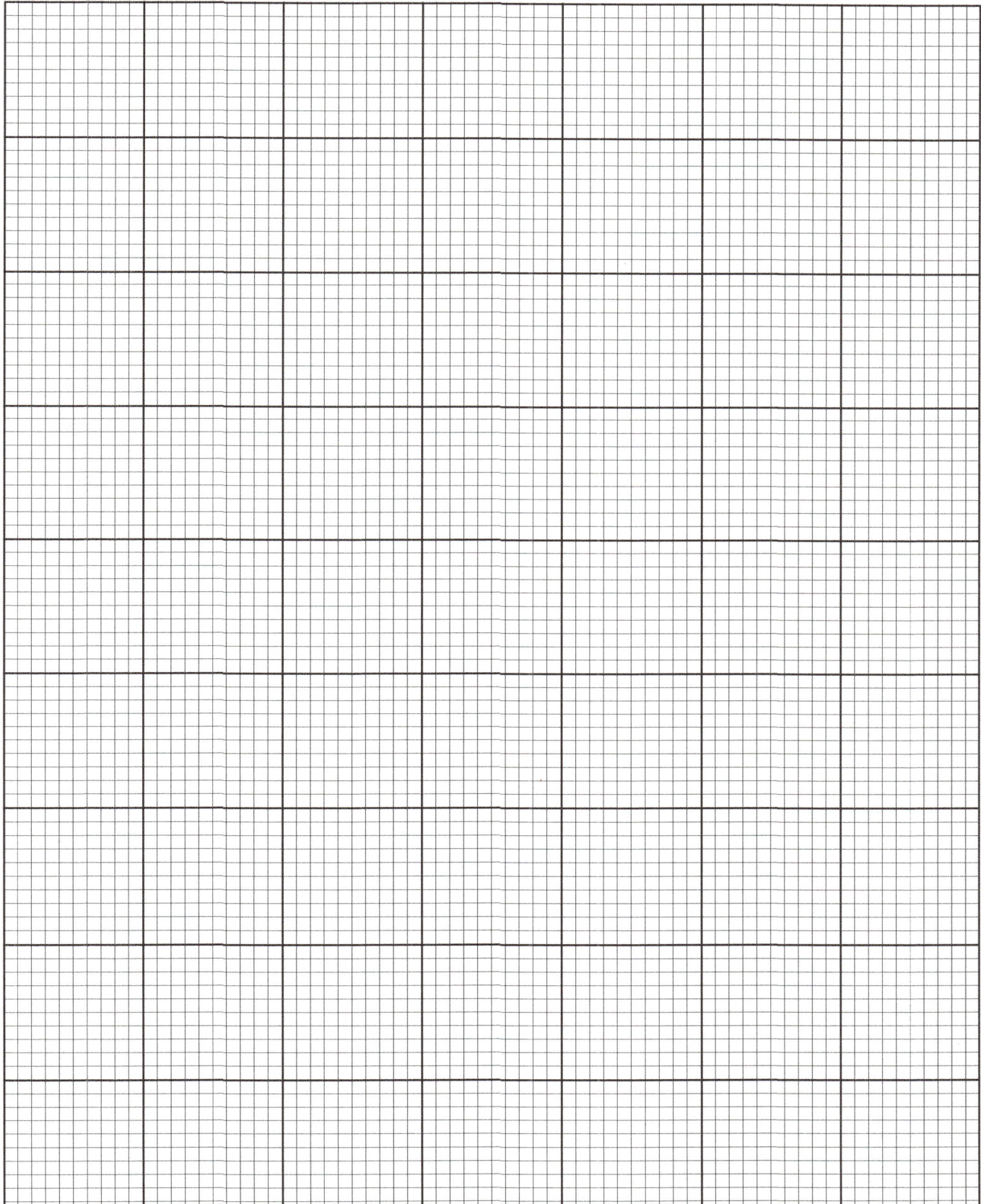

Name _____   Section _____

Soil Erosion

LAB EXERCISE 19

Equipment

- distance tapes
- stadia rod
- soil sampler
- soil sample bags

Soil erosion caused by runoff has been a major problem in the U.S. It can result in declining agricultural productivity and adversely affect water quality in nearby streams. Water erosion occurs naturally on most landscapes. However, erosion has greatly accelerated during the past century because of construction and urban development, off-road vehicles, deforestation, and a variety of agricultural practices.

In this exercise you will estimate the rate of erosion from a designated site and then determine how erosion will likely increase or decrease depending on the type of land-use changes that occur.

The **Fort Mitchell** topographic map and corresponding aerial photograph (Figure 19.1) shows an area in central Alabama with a variety of land uses. The lab instructor may instead provide a map or photograph depicting an area in your location. Select the boundaries of a site (problem area) on the map or photograph for which you will calculate the rate of erosion.

Soil erosion on the designated site will be estimated using the **universal soil-loss equation**. This soil-loss equation includes the four principal factors that affect runoff erosion: precipitation, type of soil, topography, and vegetation cover. This equation estimates soil erosion in tons per acre per year (A):

$$A = R * K * S * C$$

where: R = rainfall index
K = soil erodability factor
S = slope factor, both steepness and length
C = plant cover factor

FIGURE 19.1 Fort Mitchell, Alabama-Georgia.

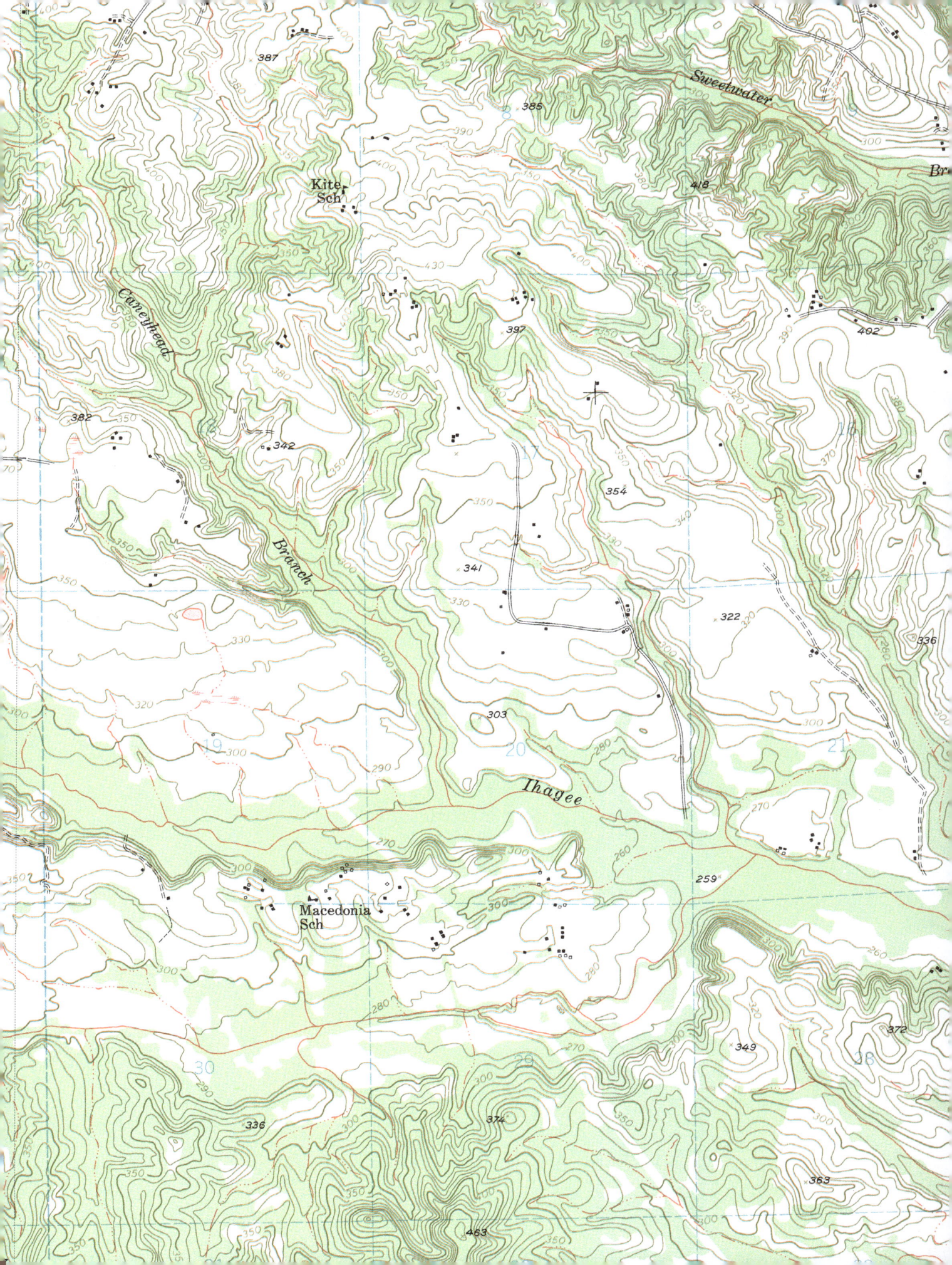

FORT MITCHELL, ALABAMA–GEORGIA

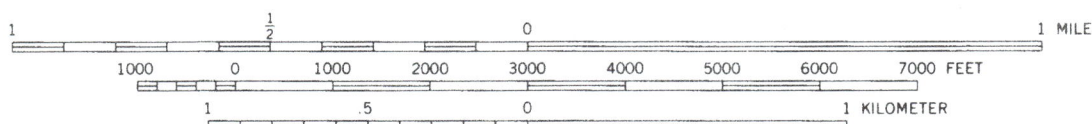

SCALE 1:24,000

Contour Interval 10 Feet

Estimate of Soil Loss

1. Select the **rainfall erosion index** for the study site.

   High intensity rainfall causes high rates of soil erosion. The incidence of high intensity storms combined with the total annual precipitation of an area were used by the Soil Conservation Service to develop a **rainfall erosion index**.

   Select the rainfall erosion index for the study area from Figure 19.2 and enter below.

   Rainfall Erosion Index (R) = _____

2. Assign a **soil erodability factor (K)**.

   This factor (K) indicates the susceptibility of the soil to sheet and rill erosion by water and is based on several soil properties (silt/fine sand and organic matter content, structure, permeability).

   The factor (K) values range from 0.05 to 0.69. The higher the value the more susceptible the soil is to water erosion. These values can be obtained from the Soil Conservation Service and are usually listed in County Soil Surveys. If K is not available, use Table 19.1 to select a value based on soil texture:

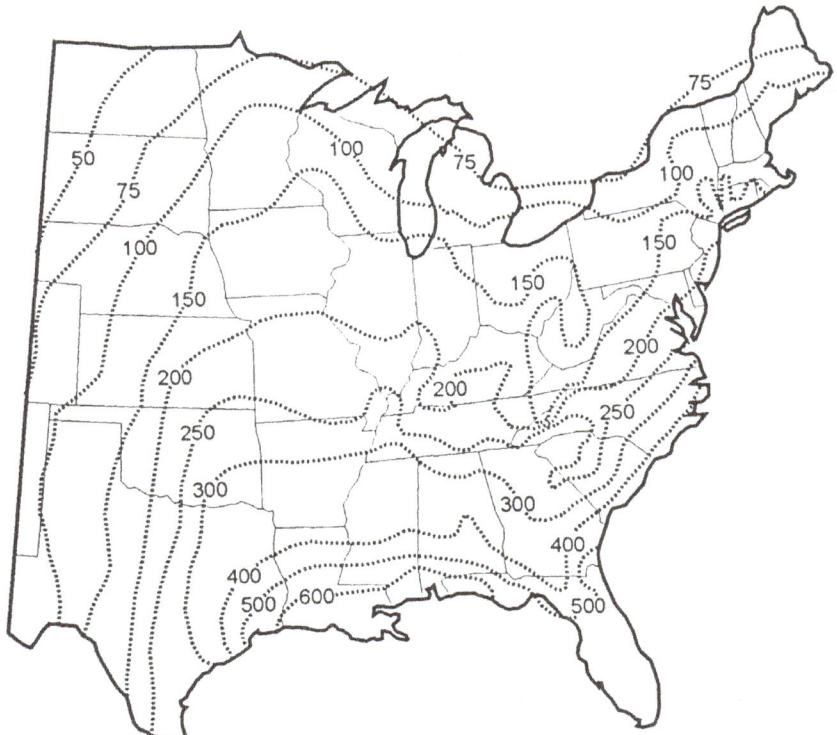

FIGURE 19.2 Rainfall erosion index map.

TABLE 19.1 SOIL ERODABILITY FACTOR (K)

clay or clay loam	0.2
silt-clay loam	0.3
silt loam	0.3
sandy loam	0.25
gravelly loam	0.2

Soil Erodability Factor (K) = _____

3. Determine the slope factor (S).

   The ability of rainfall to erode a surface increases with an increase in either the length or gradient of the slope. Both of these factors directly affect the velocity of flow downslope. Based on the topographic map, determine the slope factor to be used for the study site.

   a. Calculate the percent grade: _____ percent

   b. Determine the length of the slope: _____ feet

   Using these slope characteristics refer to Table 19.2 to determine the slope factor (S).

   Slope Factor (S) = _____

TABLE 19.2 SLOPE FACTOR BASED ON STEEPNESS AND LENGTH

Slope length (feet)	Slope Steepness (percent)								
	4	6	8	10	12	14	16	18	20
100	0.4	0.7	1.0	1.4	1.8	2.3	2.8	3.4	4.2
200	0.6	0.9	1.4	1.9	2.6	3.3	4.1	4.8	5.9
300	0.7	1.2	1.7	2.4	3.1	4.0	5.0	5.9	7.2
400	0.8	1.3	2.0	2.7	3.6	4.6	5.7	6.8	8.3
500	0.9	1.5	2.2	3.1	4.0	5.2	6.4	7.6	9.3
600	1.0	1.6	2.4	3.3	4.4	5.7	7.0	8.3	10.2
700	1.1	1.8	2.6	3.6	4.8	6.1	7.6	9.0	11.1
800	1.2	1.9	2.8	3.8	5.1	6.5	8.1	9.6	11.8
1000+	1.3	2.1	3.1	4.3	5.7	7.3	9.1	10.8	13.2

4. Determine the plant cover factor (C).

   Plant cover provides significant protection from soil erosion. Foliage intercepts rainfall, reducing the impact of raindrops and slowing the rate at which water reaches the ground surface. Also, plant roots can effectively hold soil particles in place.

   Use the aerial photograph to estimate:

   a. Canopy cover (percentage of the site covered by trees or shrubs):

   _____ percent

b. Ground cover (percentage of the site covered by grass and other low lying plants):

_____ percent

Using these vegetation cover values, refer to Table 19.3 to determine the plant cover factor (C).

Plant Cover Factor (C) = _____

TABLE 19.3 PLANT COVER FACTORS

		\multicolumn{6}{c}{Percent Groundcover}					
		0	20	40	60	80	100
	0	.45	.20	.10	.042	.013	.003
		.45	.20	.15	.090	.043	.011
	25	.39	.18	.09	.039	.013	.003
Percent		.39	.22	.14	.085	.042	.011
Canopy							
Cover	**50**	.39	.16	.08	.038	.012	.003
		.39	.19	.13	.080	.040	.011
	75	.27	.10	.08	.035	.012	.003
		.32	.18	.12	.080	.040	.011

Use the top value if the ground surface is covered with grass or organic material at least 2 inches deep. Use the bottom value if the surface is sparsely covered with weeds and there is little organic matter.

Multiply the soil erodability factor (K), slope factor (S), and the plant cover factor (C), times the Rainfall Index (R) to determine soil loss (A):

A = (R) _____ × (K) _____ × (S) _____ × (C) _____

The soil loss on this site: _____ tons/acre/year

Predict Future Soil Loss

The previous calculation of soil loss was based on conditions that occurred when the photograph was taken. The rate of erosion, however, can be changed by altering slope steepness/length, and vegetation cover, or by changes in land-use.

1. Assume the site is left undisturbed so that a forest stand develops. As the canopy develops it shades the understory and the ground cover is reduced. Predict soil-loss on this site assuming:

   canopy cover _____ percent

   ground cover _____ percent

   In this case, the soil erodability factor (K) and slope factor (S) do not change.

   The soil loss on this site: _____ tons/acre/year

   Compared to the previous problem, the (increase, decrease) in soil-loss is:

   _____ tons/acre/year

   _____ percent

   Explain why changes in vegetation will affect the rate of soil loss as shown in this problem.

2. Assume the site is cleared for cultivation or construction. In this case, canopy cover and ground cover values are 0 percent. The soil erodability factor (K) and slope factor (S) do not change.

   The soil loss on this site: _____ tons/acre/year

   Compared to the first problem, the (increase, decrease) in soil-loss is:

   _____ tons/acre/year

   _____ percent

   Explain why soil erosion is higher when vegetation is removed as shown in this problem.

Estimate Soil Loss

Your lab instructor will select a study site for you to visit to calculate soil loss.

1. Use Figure 19.2 to determine the rainfall erosion index (R) = _____

2. Assign a soil erodability factor (K). These values can be obtained from the County Soil Survey or may be provided by your lab instructor. If K is not available, use Table 19.1 to select a value based on soil texture.

    Soil Erodability Factor (K) = _____

3. Determine the slope factor (S)

    a. Calculate the percent grade: _____ percent

    b. Determine the length of the slope: _____ feet

    Using these slope characteristics refer to Table 19.2 to determine the slope factor (S). To determine percent grade and length of the slope, use equipment provided by your lab instructor. If no equipment is available, you can make fairly reliable estimates by pacing and sighting on points upslope.

    Slope Factor (S) = _____

4. Determine the plant cover factor (C)

    Visually estimate:

    a. canopy cover (percentage of the site covered by trees and shrubs):

        _____ percent

    b. ground cover (percentage of the site covered by grass and low lying plants):

        _____ percent

    Refer to Table 19.3 to determine the plant cover factor (C).

    Plant Cover Factor (C) = _____

Soil loss (A) = (R) _____ × (K) _____ × (S) _____ × (C) _____

The soil loss on this site: _____ tons/acre/year

Gully Development

Soil erosion is a normal and expected condition on almost all natural surfaces. However, the rates of erosion are usually small unless a site has been disturbed by removal of vegetation or changes in hydrologic conditions. If water is channeled, its velocity increases and gully development is likely to occur. The purpose of this exercise is to determine the amount and rate of soil erosion in a gully.

Your lab instructor will select an area in which water has been channelized and as a result a gully has developed.

Calculate the Volume of the Gully

1. Divide the gully into a few sections as shown in Figure 19.3. The boundaries between each section should be _____ feet apart (equidistant). At each line separating sections measure (in feet):

    a. the width at the top and bottom of the gully

    b. the depth (at least three measures of depth that are equidistant apart)

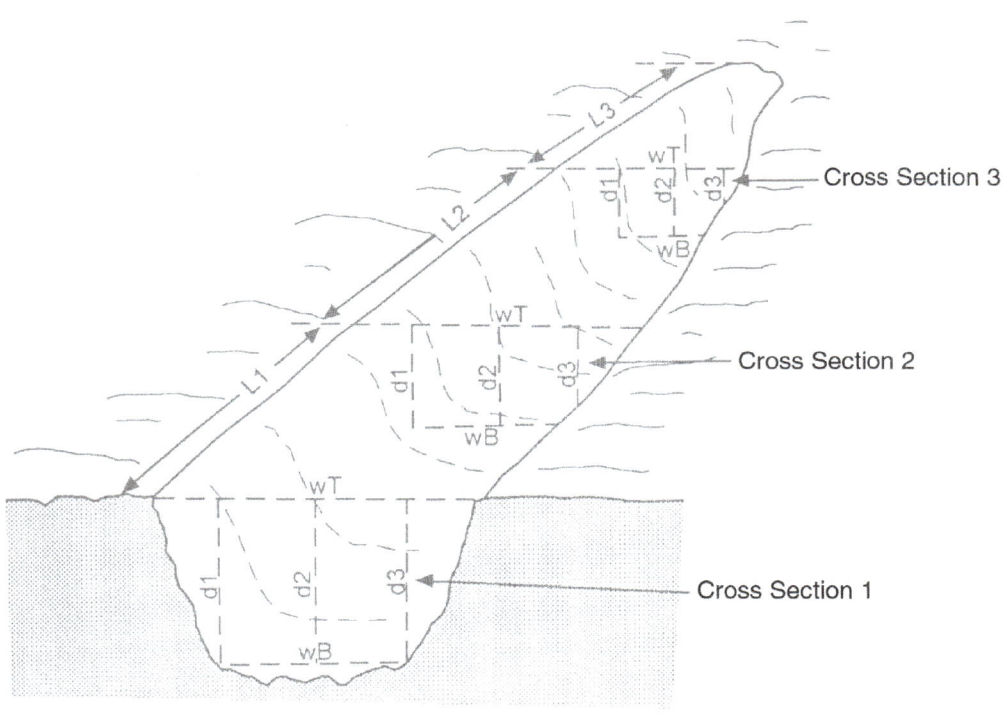

d   Depth
L   Length
wT Width across top
wB Width across bottom

FIGURE 19.3 Calculate the volume of the gully.

Enter these values below:

Section 1: width top _____ depth _____
            width bottom _____ depth _____
                                     depth _____
                                     depth _____

Section 2: width top _____ depth _____
            width bottom _____ depth _____
                                     depth _____
                                     depth _____

Section 3: width top _____ depth _____
            width bottom _____ depth _____
                                     depth _____
                                     depth _____

2. Determine the cross-sectional area for each section boundary.

   Find the *average* for the depth measurements (feet) and multiply by the average of the widths (feet)(top and bottom).

   Section 1:

   depth _____ × _____ width

   = cross-sectional area (cs1) _____ feet^2

   Section 2:

   depth _____ × _____ width

   = cross-sectional area (cs2) _____ feet^2

   Section 3:

   depth _____ × _____ width

   = cross-sectional area (cs3) _____ feet^2

3. Determine the volume of the gully:

   Average the cross-sectional area (cs) between consecutive section boundaries and multiply by the length between each to determine the volume of that section of the gully.

   volume of the first gully segment = _____ feet^3

   volume of the second gully segment = _____ feet^3

   volume of the third gully segment = _____ feet^3

   Add the volumes of the gully segments to determine the total volume of soil loss.

   _____ feet^3

Determine the Weight of Soil Loss

The class will be divided into groups for soil sample collection. Extract an undisturbed soil sample that will fit in a graduated cylinder. Be sure not to compact the soil sample or allow it to break apart. Record its volume:

_____ cm^3 (ml)

Dry the soil to determine its weight: _____ g

Convert the weight of this soil sample to lbs/feet^3, using the following conversions:

   1 kg (1000g) = 2.204 lbs
   1 cm^3 (ml) = 0.061 inches^3

   (dry weight) _____ lbs/(volume) _____ feet^3

If you were unable to obtain a soil sample to determine its weight, use the following values:

clay   80 lbs/ft^3
clay or silt loams   100 lbs/ft^3
sand or sandy loams   120 lbs/ft^3

Determine the total weight of the soil lost from the gully. (1 ton = 2000 lbs)

   _____ lbs/feet^3 × _____ total soil loss (feet^3)

   soil loss: _____ tons

   Years since the gully formed: _____

   Average soil loss per year: _____ tons

The average rate of soil loss on cultivated land in the U.S. is about 4 tons/acre/year. Soil loss greater than about 5 tons/acre/year is considered serious and could result in declining agricultural productivity. How does the rate of erosion for the gully you measured compare to that of the U.S.?

Name _____ Section _____

Runoff and Infiltration

LAB EXERCISE
20

Most precipitation will either infiltrate the soil surface or runoff, and only a small proportion will evaporate. The purpose of this exercise is (1) to determine for a specific area the amount of rainfall that runs-off and eventually enters streams, or that infiltrates the ground surface, and (2) to determine how human impact can alter the runoff/infiltration ratio.

Water can readily infiltrate most dry soils. As water enters the soil surface the pore spaces in the soils become filled. The rate of infiltration slows and much of the additional precipitation will run off. On naturally vegetated surfaces there is little runoff except during very heavy periods of precipitation. In contrast, paved surfaces and those occupied by buildings will have no infiltration, and therefore, all precipitation runs-off.

Figure 20.1 is an aerial photograph showing the suburban edge of Huntsville, Alabama. The built-up areas are mostly residential. The lab instructor will select a portion of this photograph or may provide a photograph showing an area in your location for which you will determine the runoff/infiltration ratio.

The R.F. of the photograph is 1: _____.

First, calculate the designated area: _____ m^2

A standard area measurement is a hectare (ha) which is 100×100 meters, or $10,000 m^2$. Convert the designated area to hectares.

_____ ha

Next, estimate the percentage of the drainage basin that is paved. Typically, about one-half of the surface occupied by single family housing is paved. The percentage of the surface paved is usually much higher for other types of land-use (commercial, industrial).

Percentage of the surface that is paved: _____

No infiltration occurs on paved surfaces, and therefore, runoff is 100 percent of precipitation.

On vegetated or non-paved surfaces precipitation will infiltrate the ground surface, but the amount will vary depending in large part on the amount of water already in the soil and different soil types. If there is very little precipitation no runoff will occur. However, after water enters the soil surface the infiltration rate slows because many of the pore spaces in the surface soil layer are now filled with water.

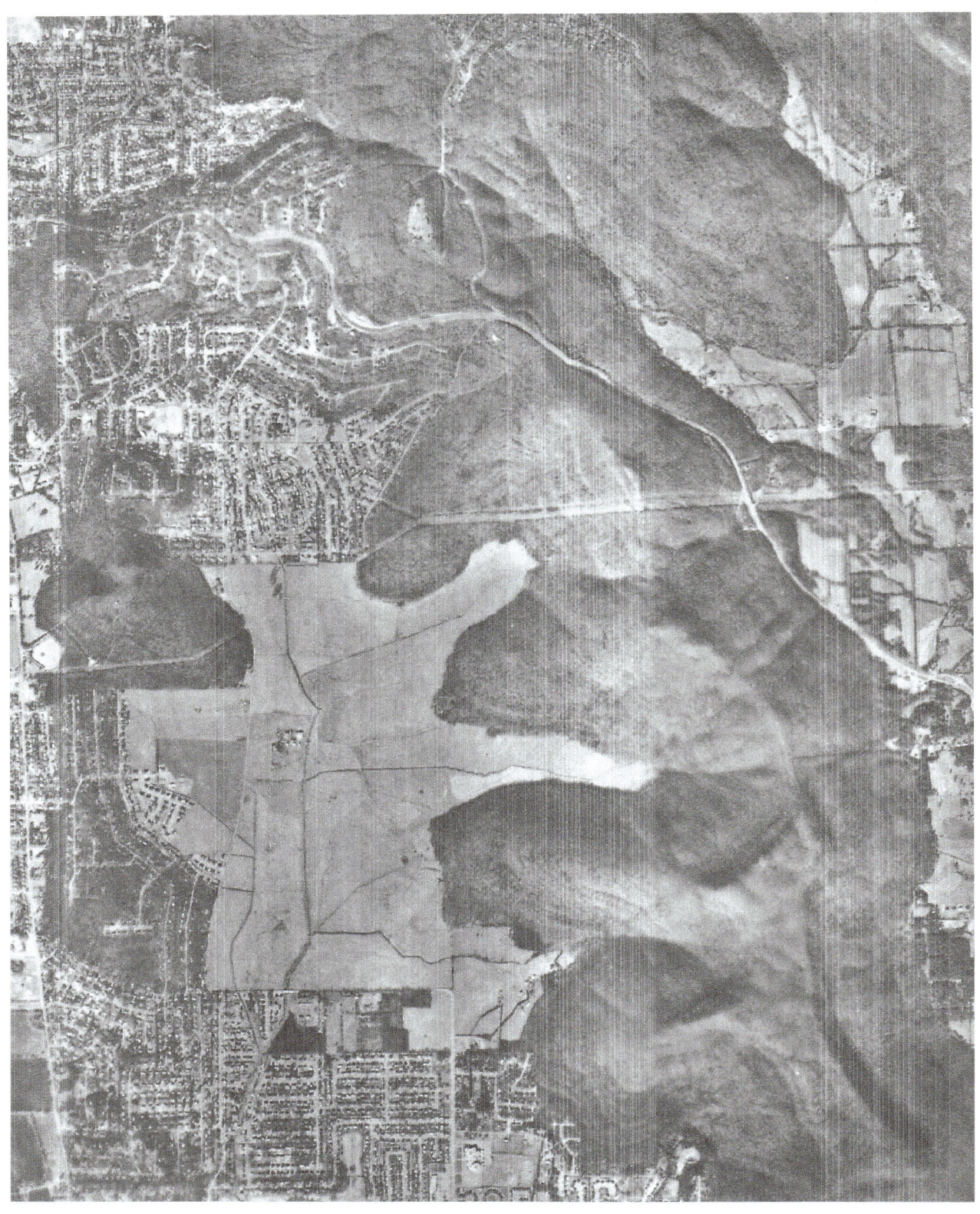

FIGURE 20.1 Suburban edge of Huntsville, Alabama.

The infiltration rate also depends on soil texture. Sandy soils have large pore spaces that allow water to move easily through them. The finer texture soils (silt and clay) have smaller pore sizes that restrict water movement, and therefore, infiltration is slower.

Use the following table (20.1) to determine rates of infiltration for different soil textural classes during (1) the first 30 minutes of precipitation, and (2) for the remaining period.

TABLE 20.1 RATE OF INFILTRATION ON VEGETATED (NON-PAVED) SURFACES

Soil Textural Class	Rate of Infiltration (cm/hour)	
	first 30 mins.	remaining period
clay or clay loam	0.4	0.1
silt-clay loam	1.0	0.2
silt loam	1.2	0.3
sandy loam	2.0	0.4

Soil Textural Class _____

Infiltration   a. first 30 mins. _____ (cm/hour)

                  b. remaining period _____ (cm/hour)

Suppose 6 cm of rain falls on the designated area in a 24 hour period.

1. What is the average rainfall intensity in cm per hour?

    _____

2. Assuming this intensity to hold true during the first 30 minutes, how much runoff would occur?

    from the paved area _____ cm

    from the vegetated area _____ cm

3. After 30 minutes how much runoff would occur per hour?

    from the paved area _____ cm

    from the vegetated area _____ cm

4. How much runoff would occur in the entire 24 hour period?

    from the paved area _____ cm

    from the vegetated area _____ cm

Use the following equation to determine the volume of runoff (V) in cubic meters:

$$V = (A * R)/100$$

where: A = the area (m²)
R = the runoff (cm)

Note: The above value (A * R) is divided by 100. Remember, there are 100 cm in a meter, and this allows V to be measured in cubic meters.

5. How many cubic meters ran off this site in the 24 hour period?

   _____

6. What percentage of the total rain that fell ran off?

   _____ percent

The runoff/infiltration ratio for the same area changes as land-use changes. For this problem assume 6 cm of rain falls on the same site in a 24 hour period, as in previous problem. However, more buildings and parking lots are constructed. Therefore, the total paved area increases and vegetated area decreases.

   Assume that the paved area increases _____ percent, or _____ m².

   What is the paved area? _____ m²

   What percentage of the drainage basin is now paved? _____
   (= paved area/designated area)

   vegetated? _____

1. What is the average rainfall intensity in cm per hour?

   _____

2. Assuming this intensity to hold true during the first 30 minutes, how much runoff would occur?

   from the paved area _____ cm

   from the vegetated area _____ cm

3. After 30 minutes how much runoff would occur per hour?

   from the paved area _____ cm

   from the vegetated area _____ cm

4. How much runoff would occur in the entire 24 hour period?

   from the paved area _____ cm

   from the vegetated area _____ cm

5. How many cubic meters would run off this site in the 24 hour period?

   _____

6. What percentage of the total rain that fell ran off?

   _____ percent

7. What is the percent change in runoff compared to that in the first problem in this exercise?

   _____ percent (increase, decrease)

For the following problem assume _____ cm of rain falls in a _____ hour period.

$$\text{Designated area } \underline{\hspace{2cm}} \text{ m}^2$$

$$\underline{\hspace{2cm}} \text{ ha}$$

Percentage of the drainage basin that is paved _____

paved area: _____ m²

vegetated area: _____ m²

Soil Textural Class _____

Infiltration   a. first 30 mins. _____ (cm/hour)

                b. remaining period _____ (cm/hour)

1. What is the average rainfall intensity in cm per hour?

   _____

2. Assuming this intensity to hold true during the first 30 minutes, how much runoff would occur?

   from the paved area _____ cm

   from the vegetated area _____ cm

3. After 30 minutes how much runoff would occur per hour?

   from the paved area _____ cm

   from the vegetated area _____ cm

4. How much runoff would occur in the entire 24 hour period?

   from the paved area _____ cm

   from the vegetated area _____ cm

5. How many cubic meters would run off this site in the 24 hour period?

   _____

6. What percentage of the total rain that fell ran off?

   _____ percent

7. What is the percent change in runoff compared to that in the first problem in this exercise?

   _____ percent (increase, decrease)

Name_____ Section_____

LAB EXERCISE 21

Stream Discharge

Stream discharge is the amount of water flowing through a channel. Discharge directly affects velocity which determines a stream's ability to erode its channel and transport sediment. In this exercise you will (1) observe a small stream to calculate its discharge, (2) estimate velocity and discharge for streams that cannot be directly measured, and (3) predict average discharge based on the size of the drainage area and precipitation.

Calculate Discharge

Your lab instructor will select a small stream to be used in this exercise. Discharge (Q) is calculated by:

$$Q = AV$$

where: A = cross-sectional area of the channel, and
V = average velocity.

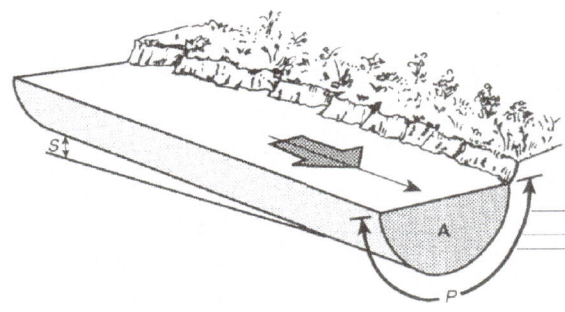

FIGURE 21.1 Cross-section (A), perimeter (P) and slope (S) of a stream channel.

1. Determine the cross-sectional area of the channel.

   Measure the width of the channel.

   width = _____ m

237

Measure the water depth. If the depth varies significantly, you should make several measurements and calculate the average.

$$\text{depth} = \underline{\hspace{3cm}} \text{ m}$$

Calculate the cross-sectional area by multiplying width × water depth.

$$\text{cross-sectional area} = \underline{\hspace{3cm}} \text{ m}^2$$

2. Determine the stream velocity.

   Calculate the average velocity, which is expressed as m/second. The velocity can be determined by timing the passage of a float along a known distance. For example, if your float takes 10 seconds to go 20 meters, then 20m/10 = 2 m/second.

   Repeat the velocity measurement 5 times. Complete the chart below with your observations:

	Observations				
	1	*2*	*3*	*4*	*5*
Distance (m)	_____	_____	_____	_____	_____
Time (sec)	_____	_____	_____	_____	_____
Velocity (m/sec)	_____	_____	_____	_____	_____

   $$\text{Average velocity} = \underline{\hspace{3cm}} \text{ m/sec}$$

   Now, multiply A (the cross-sectional area) × V (average velocity) to calculate Q (discharge).

   $$Q = \underline{\hspace{3cm}} \text{ m}^3/\text{second}$$

   Based on this discharge, what is the daily volume of water that flows through this stream?

   $$\underline{\hspace{3cm}} \text{ m}^3$$

Determine the Erosion Potential

The velocity of water determines its ability to move or transport sediment. Based on the stream velocity you measured, determine the largest particle size that can be moved by referring to Figure 21.2 shown on page 239.

Maximum particle size that can be transported _____ mm

   Textural class (clay, silt, etc.) _____

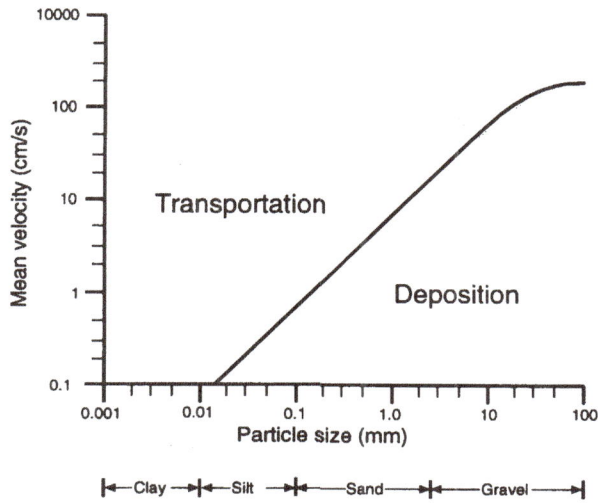

FIGURE 21.2   Relationship between water velocity and particle size that can be transported.

Estimation of Bankfull Discharge

Bankfull refers to the channel being completely filled with water. Any additional water will cause flooding because it cannot be held in the channel and is forced into the surrounding floodplain. Bankfull discharge can be easily calculated.

Draw a cross-section of the stream channel as accurately as possible on Figure 21.3. Then, determine the cross-sectional area for the channel (assuming the channel is completely full).

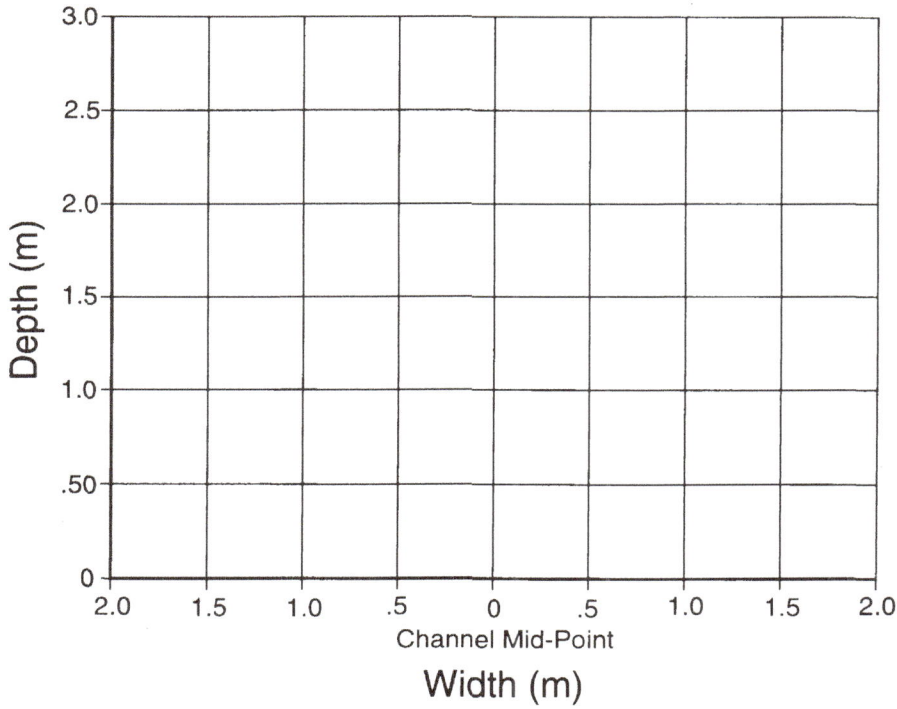

FIGURE 21.3   Cross-section of stream channel.

Bankfull A (cross-sectional area) = _____ m²

You must know the water velocity to estimate the bankfull discharge. In this case you do not know the velocity, because as Q increases, so does V. But in this problem assume that V = _____ m/sec.

1. Based on the channel cross-sectional area and the estimated velocity, the Q (discharge) at bankfull is

   _____ m³/sec

2. Based on this discharge, what is the daily volume of water that flows through this stream if the stage is bankfull?

   _____ m³

3. What is the percentage increase of this value compared to observed discharge?

   _____ percent

4. Based on the estimated stream velocity at bankfull, what is the largest particle size that can be moved? (Refer to Figure 21.2.)

   _____ mm

   Textural class (clay, silt, etc.): _____

Estimating Stream Velocity

The velocity of a stream can be estimated with reasonable accuracy when direct measurements are not possible. Velocity is controlled by the three following channel characteristics:

1. hydraulic radius (which is a measure of the proportion of water in a stream that comes in contact with the channel bed or banks)
2. channel slope or gradient
3. roughness of the channel

Stream velocity (V) can be estimated using Manning's formula which takes into account these three channel characteristics:

$$V = \frac{1.0}{n} * R^{.67} * S^{.5}$$

where: V = in m/second
R = hydraulic radius (the cross-sectional area of the channel divided by the wetted perimeter of the channel)
S = channel slope or gradient
n = roughness coefficient

The lab instructor will select a stream shown on a topographic map. You will estimate stream velocity at (1) average flow and (2) bankfull, using Manning's formula. A worksheet to help calculate velocity is included.

The topographic map _____ has an R.F. of 1: _____

Estimate velocity for the following stream: _____
during (1) average flow and (2) at bankfull.

Enter the values on the blank lines and on the worksheet that follows.

1. Calculate the Hydraulic Radius (R)

    Larger, deeper streams generally flow faster than smaller streams that have the same gradient. The greater the discharge (and deeper the channel) the smaller the proportion of water in the channel that comes in contact with the channel banks, and is therefore slowed by friction. The hydraulic radius (R) is a value representing the ratio of (1) the cross-section of the channel (as shown in Figure 21.1) and (2) the wetted perimeter, or length of the channel bed plus the height of the banks occupied by water.

    Use the topographic map to determine the width of the channel. You cannot determine the water depth from a topographic map. In this problem assume that the average water depth at normal flow is _____ percent of the channel width.

    depth at normal flow: _____

    Based on the depth at average flow (that is depicted on the topographic map) and the elevation of the floodplain adjacent to the channel, determine:

    depth at bankfull: _____

    Draw the cross-section of the stream at average flow and bankfull below on Figure 21.4.

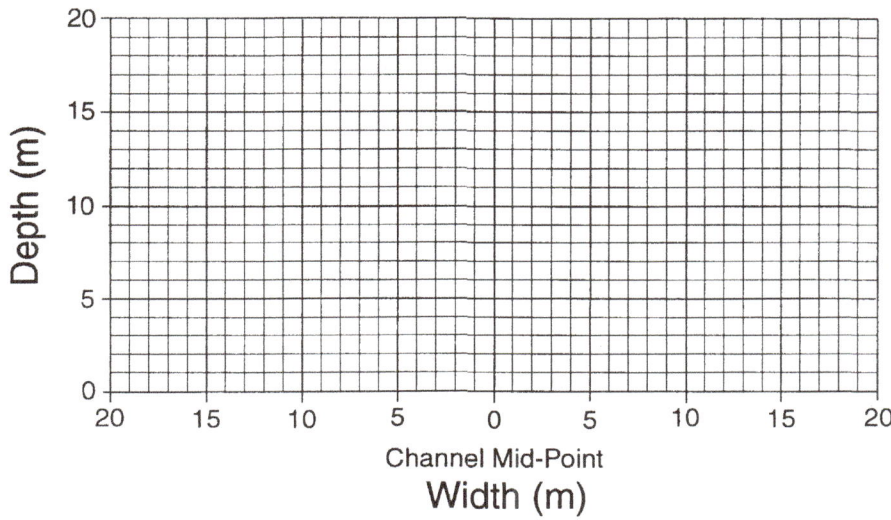

FIGURE 21.4 Channel cross-section of the stream at average flow and bankfull.

To calculate the cross-sectional area determine:

a. channel width: _____ m

   channel width (bkf): _____ m

b. channel depth: _____ m

   channel depth (bkf): _____ m

Calculate the channel cross-sectional area (w * d):

   average flow: _____ m²

   bankfull: _____ m²

Calculate the wetted perimeter (w + 2d):

   average flow: _____ m

   bankfull: _____ m

Hydraulic radius (R) (cross-sectional area/wetted perimeter):

   average flow: _____

   bankfull: _____

2. Calculate Stream Gradient (S)

Remember that the stream gradient is a ratio of the elevational difference between two points along the stream and the channel length (or distance) between them. The steeper the gradient, the faster the water will move.

$$S = \frac{rise}{run}$$

rise = the elevational difference between two points on the channel bed

_____ ft or m

run = the distance between these points.

_____ ft or m

Slope or stream gradient (S): _____

3. Determine Roughness Coefficient (n)

Water moves fastest through a smooth, straight channel. Irregular channels (those that meander and have channel obstructions) have a much slower velocity. Channel roughness is a measure of channel characteristics that directly affect velocity including: channel configuration (straight, meandering, etc.), the presence of vegetation or large boulders, and if the channel sides and bottom are smooth or irregular. As channel roughness increases, flow velocity decreases.

Approximate values for n are:

Alluvial stream:
    straight, clean channels   0.030
    straight, weedy or with boulders   0.035
    clean winding channels   0.040

Mountain streams:
    cobbles and boulders   0.050
    gravel bottom   0.030
    sand bottom   0.020

Roughness coefficient (n) _____

WORKSHEET

	Normal Flow	Bankfull
width:	_____	_____
depth:	_____	_____
cross-sectional area:	_____	_____
wetted perimeter:	_____	_____
Hydraulic radius (R):	_____	_____
Stream gradient (S):	_____	_____
Roughness coefficient (n):	_____	_____

Enter the values for the hydraulic radius (R), stream gradient (S), and the roughness coefficient (n) into Manning's formula to estimate stream velocity.

              normal flow   *bankfull*

velocity: _____ m/sec   _____ m/sec

discharge: _____ m³/sec   _____ m³/sec

Based on this discharge, calculate the daily volume of water that flows through this stream.

daily discharge: _____ m³   _____ m³

Determine the largest particle size that can be moved. (Refer to Figure 21.2.)

particle size: _____ mm   _____ mm

textural class: _____   _____

1. If your calculations are correct, stream velocity is greater at bankfull than normal (average) flow. Why does the water move faster when the water level is higher?

2. Based on the relationship between velocity and the particle size that can be transported, what are the effects of floods compared to average flows on erosion and moving channel sediment?

Predicting Stream Discharge

In the previous problems in this exercise you have estimated discharge by directly measuring channel dimensions and velocity, and later by using topographic maps to indirectly determine these same stream conditions. The average discharge of a stream can be roughly predicted by using the following equation that takes into account the area of the drainage basin and precipitation:

$$Q = 0.082 * A^{1.02} * P^{0.75}$$

where: Q = mean annual discharge (ft³/second)
 A = basin area (square miles)
 P = mean annual precipitation (inches) minus 30

The Heidelberg topographic map depicts an area in central Kentucky. Predict the mean annual discharge for Cave Branch, a small stream that is a tributary of the Kentucky River. The lab instructor may instead provide a map or aerial photograph of a site closer to your location to determine discharge.

Mean annual precipitation in this drainage basin is _____ inches.

1. Determine the area of the drainage basin: _____ miles²

2. Use these values in the above equation to predict the mean annual discharge.

    _____ ft³/sec

3. Based on predicted discharge, what is the daily volume of water that flows through this stream?

    _____ ft³

HEIDELBERG, KENTUCKY

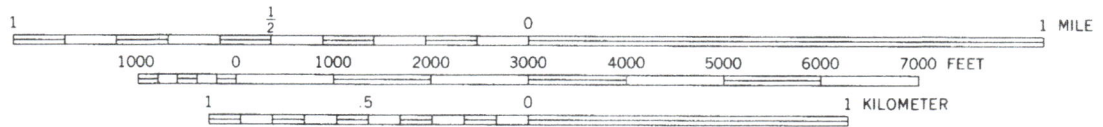

SCALE 1:24,000

Contour Interval 20 Feet

4. The Cave Branch watershed is mostly forested and only a small portion of its surface area has been cleared. The use of this equation to determine average discharge does not take into account land-use patterns. How will discharge be affected if a large section of the drainage that is currently covered by natural vegetation is cleared for agricultural uses, either pasture or cultivation?

5. How will discharge be affected if a large section of the drainage that is currently covered by natural vegetation is paved?

Name _____ Section _____

Recurrence Intervals

LAB EXERCISE 22

The frequency of natural events (such as flooding, droughts, and natural fires) can be estimated by using historical records. Rather than predicting that a natural event will occur in a given year (e.g., the Mississippi River will flood in the year 2004), the technique for calculating a recurrence interval is an attempt to determine, or at least estimate the probability that an event of a certain magnitude will occur in any given year. The purpose of this exercise is to explain the method for estimating the probability of the magnitude and frequency of natural events and to recognize the limitations of this technique.

The recurrence interval is determined by:

$$RI = (n + 1)/m$$

where: RI = recurrence interval in years
n = number of years of observations
m = rank of the observation

The following table is a 13-year record (1978–1991) of annual precipitation for a climatic recording station in Colorado. Each year has been ranked so that the wettest is #1 and the driest is #13. The recurrence interval (in years) has been calculated using the above equation in an attempt to estimate the probability of this area receiving a certain amount of precipitation for any one year.

Year	Precip (in)	Rank	Recurrence Interval (yrs)
1978	22.87	5	2.8
1979	12.08	11	1.3
1981	14.83	9	1.5
1982	13.88	10	1.4
1983	19.75	7	2.0
1984	16.02	8	1.7
1985	22.68	6	2.3
1986	9.78	13	1.0
1987	30.07	1	14.0
1988	27.49	2	7.0
1989	27.31	3	4.6
1990	25.09	4	3.5
1991	10.75	12	1.1

So, in this example it has been estimated that a year with at least 25 inches of precipitation (1990) will occur once every 3.5 years. A year with at least 30.0 inches of annual precipitation (1987) will occur once every 14 years.

The amount of annual precipitation and corresponding return intervals are plotted on Figure 22.1. A line has been drawn that best fits these points.

Predictions of the frequency (recurrence interval) of a higher annual precipitation are based on continuing the line with the same slope beyond the data until it intersects the vertical axis on the right-hand side of the graph.

Based on the precipitation curve in Figure 22.1, how much annual precipitation will occur at least once every:

50 years? _____ inches

100 years? _____ inches

200 years? _____ inches

In this case only 13 years of data was used. A longer record will provide for more accurate predictions.

Probability

Based on the recurrence interval, you can estimate the probability that an event of a certain magnitude will occur in any one year. The probability (P) is the inverse of the recurrence interval for any one year, and is calculated by:

$$P = 1/RI * 100$$

Therefore, for the year 1990 the return interval is 3.5 years (for there to be at least 25 inches of precipitation). Therefore:

$$1/3.5 * 100 = 28.5$$

or, a 28.5% probability of at least that much rainfall (or an event of that magnitude) occurring for any one year.

Precipitation Recurrence Interval

A record of the highest annual rainfall events for each year between 1970 and 1989 for Baton Rouge, Louisiana, is shown below. Determine the rank and recurrence intervals (RI) for the data set. The highest rainfall year will be ranked #1 and the lowest will be ranked #20.

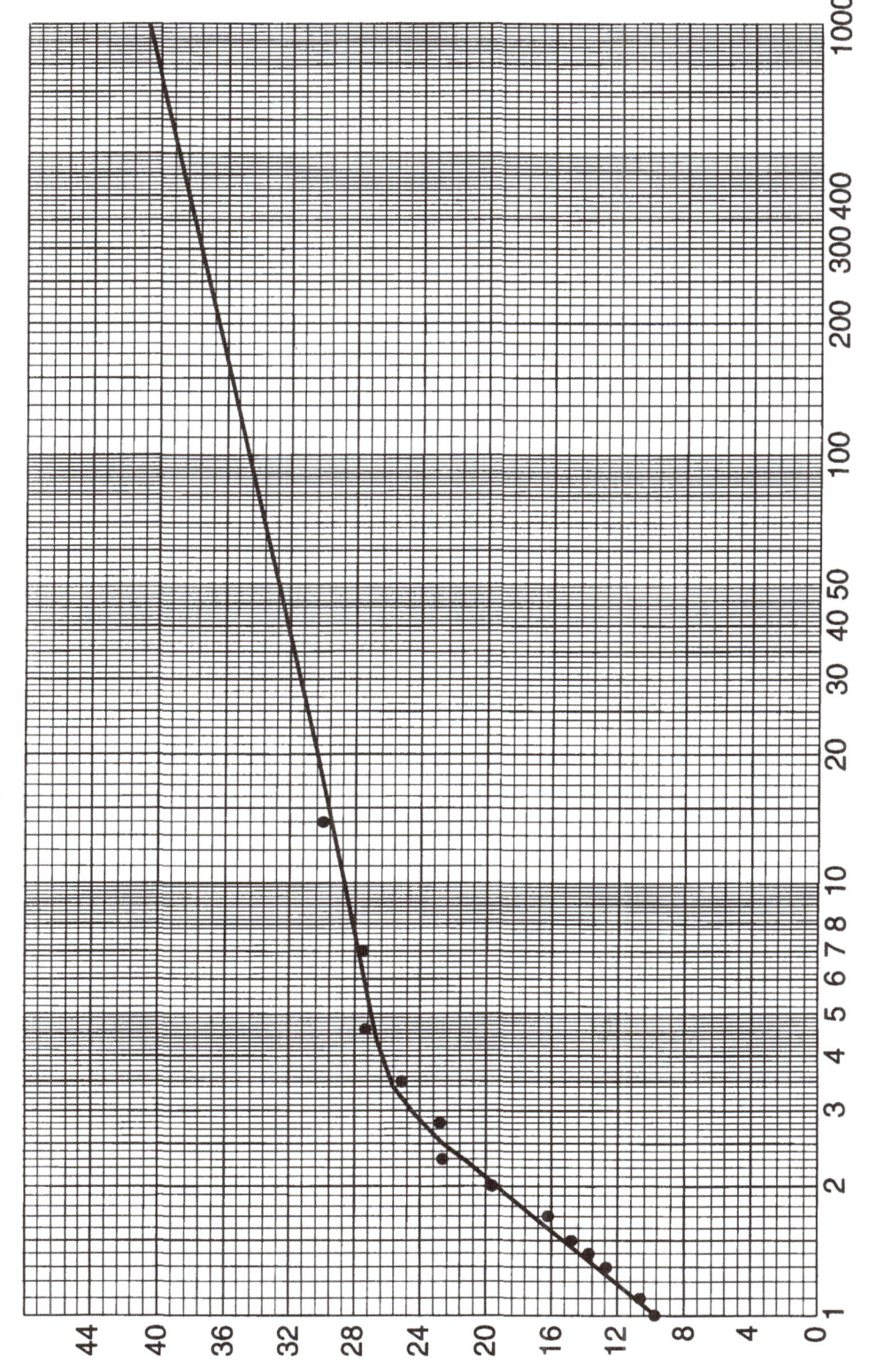

FIGURE 22.1 Precipitation curve.

Year	Maximum Rainfall	Rank	RI
1970	2.69		
1971	3.28		
1972	3.85		
1973	5.13		
1974	2.02		
1975	3.08		
1976	2.61		
1977	4.51		
1978	2.59		
1979	4.07		
1980	6.37		
1981	2.29		
1982	7.25		
1983	7.02		
1984	4.59		
1985	2.57		
1986	3.20		
1987	7.31		
1988	3.32		
1989	7.81		

Plot each of the data points on Figure 22.2.

Draw a line that best fits the data to show the overall trend. As in Figure 22.1, the line will not necessarily be straight. Although the data is for only 20 years, you can determine the 50 year (and 100 or 500 year) maximum annual rainfall events by continuing the line that fits the data (with the same slope) until it intersects the vertical axis on the right-hand side of the graph. Note that this line is not horizontal but is inclined upward.

Based on Figure 22.2 answer the following questions.

1. What is the probability that for any one year there will be at least one rainfall event no less than:

    6 inches? _____ %

    7 inches? _____ %

    8 inches? _____ %

2. What is the predicted

    50 year rainfall event? _____ inches

    100 year rainfall event? _____ inches

    500 year rainfall event? _____ inches

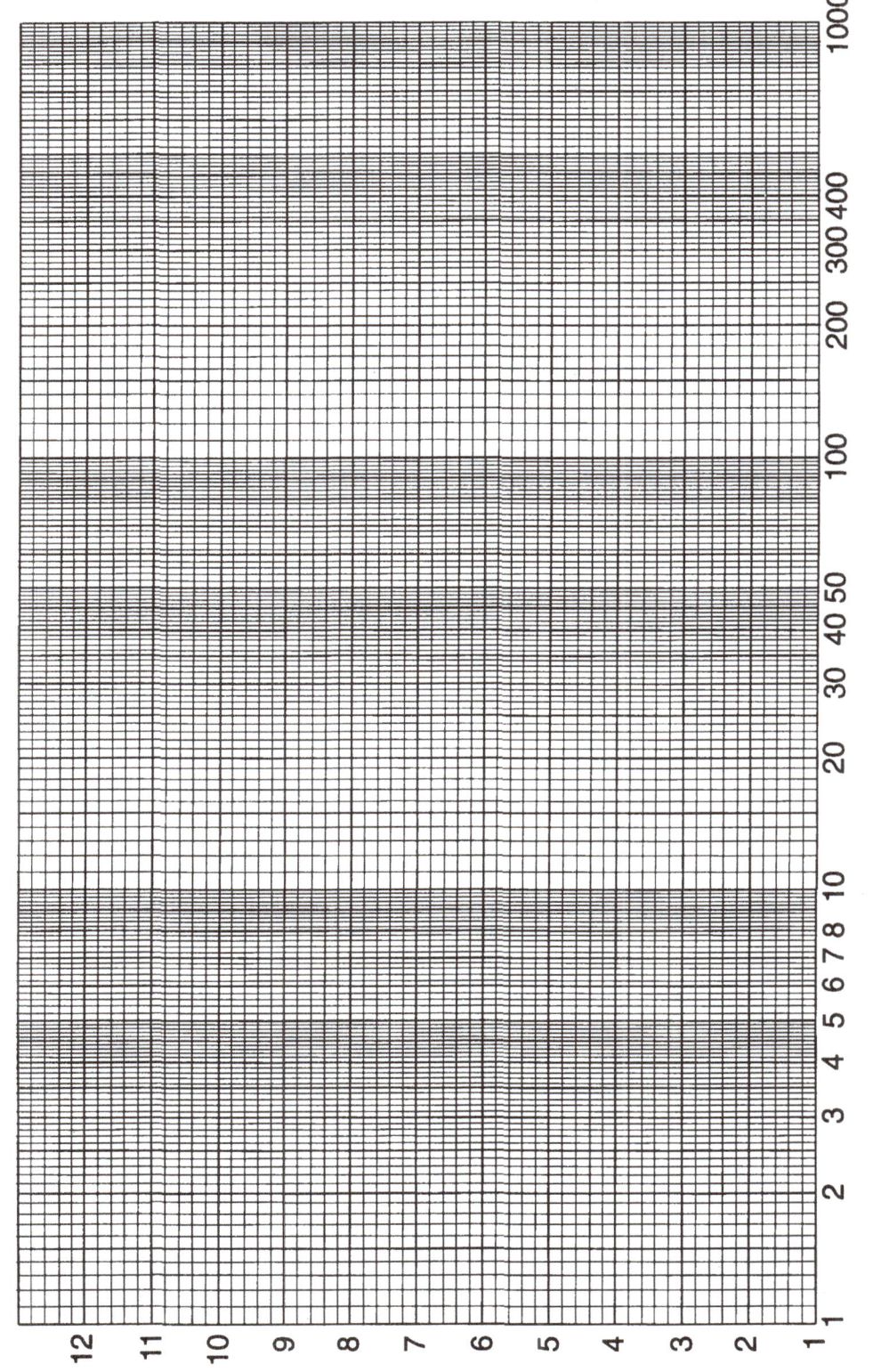

FIGURE 22.2 Precipitation curve.

3. Will the predicted 50, 100, and 500 year rainfall events necessarily occur at those time intervals? Explain.

4. Will future rainfall patterns and events reflect what has happened in the recent past? Explain.

Flood Recurrence Interval

The following table is a yearly record for the highest stream stage from the Hatchie River in western Tennessee. Stage is the height of the water, or the water surface elevation above sea level. This data can be used to determine how often a stream floods and what portion of a floodplain will be submerged.

Determine the rank and recurrence intervals (RI) for the data set. The highest stage will be ranked #1.

Year	Maximum Stage	Rank	RI
1956	256.6		
1957	257.9		
1958	254.8		
1959	254.9		
1960	254.7		
1961	256.1		
1962	257.2		
1963	253.5		
1964	255.2		
1965	258.2		
1966	255.3		
1967	255.0		
1968	255.6		
1969	257.8		
1970	255.4		
1971	255.5		
1972	255.7		
1973	259.5		
1974	256.3		

Plot each of the data points on Figure 22.3.
Draw a line that best fits the data to show the overall trend, and continue the line with the same slope until it intersects the vertical axis on the right-hand side of the graph.
Based on Figure 22.3 answer the following questions.

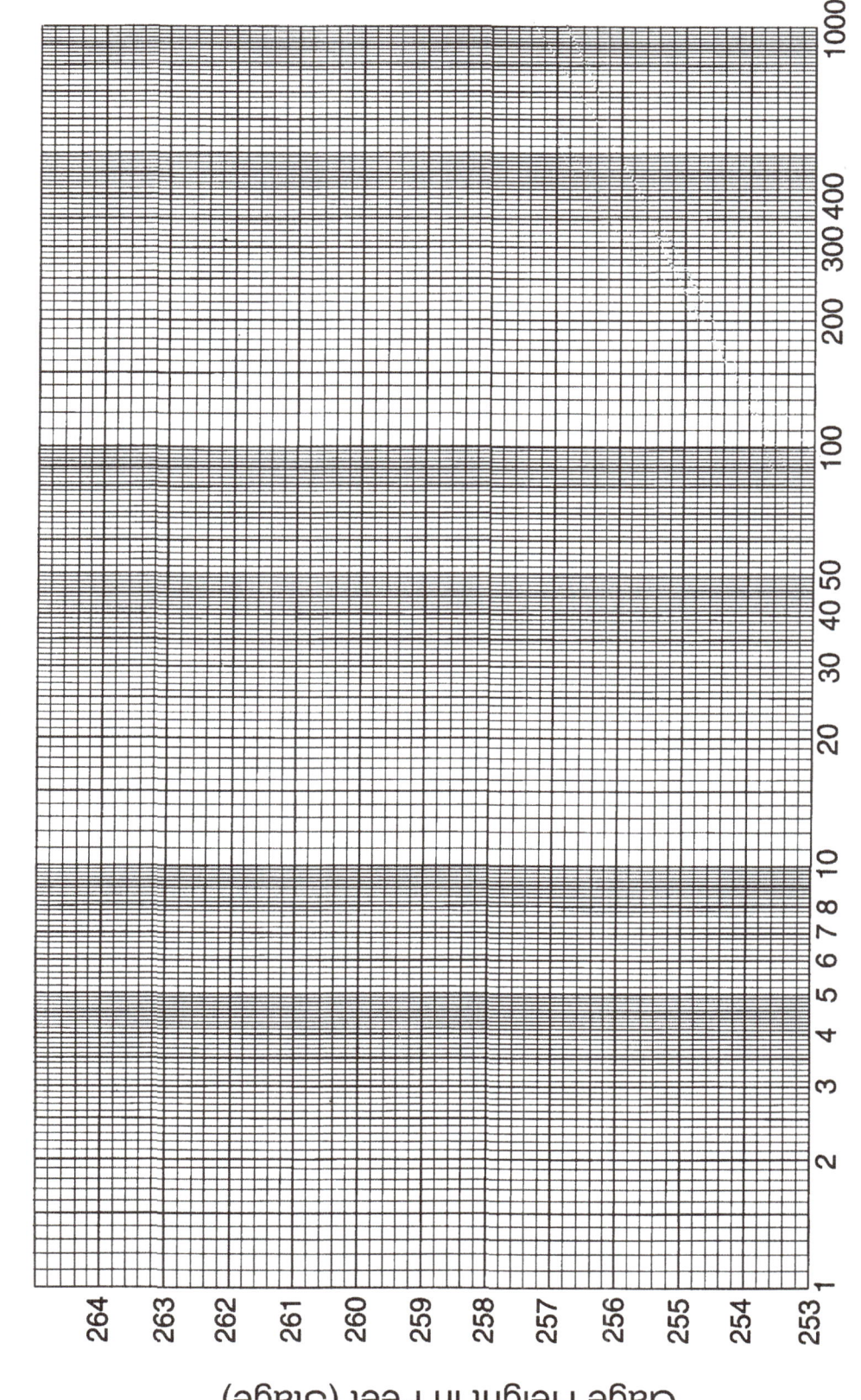

FIGURE 22.3 Flood frequency curve.

1. Flood stage is the elevation at which water is above the channel banks and therefore moving across the floodplain. Flood stage is 253 feet at this gaging station. Based on the stage record, what is the probability that for any year the Hatchie River will flood?

   _____ %

2. What is the probability for any year that stage will reach a height of 256 feet above sea level?

   _____ %

3. What is the probability for any year that stage will reach a height of 259 feet above sea level?

   _____ %

4. What is the predicted stage for the 50 year flood?

   _____ feet elevation

   Remember, you do not have 50 years of data. The 50 year flood is only a prediction. The farther you go beyond the limits of your data (100, 200, 500 year floods) the less reliable the prediction.

5. What is the predicted stage for the 100 year flood?

   _____ feet elevation

   The differences in yearly maximum stage may seem small. But, floodplains are areas of low relief, and slight increases in water level can cause large areas near the river channel to go under water.

   The South Ripley map shows a section of the Hatchie River in western Tennessee. The stage data in the table above was collected at the gaging station on the bridge crossing the river. You can determine approximately what portion of the floodplain will be submerged during a flood by matching stage data with the elevation on the map which is shown by contour lines. Contour line intervals on this map are 10 feet (and 5 feet in some areas of the floodplain).

6. Measure along the highway on the north side of the river to determine how far from the river channel the floodplain will be covered with water if stage is:

   255 feet: _____

   260 feet: _____

7. How far from the river channel will the floodplain be covered with water during the 50 year flood measuring along the highway?

   _____

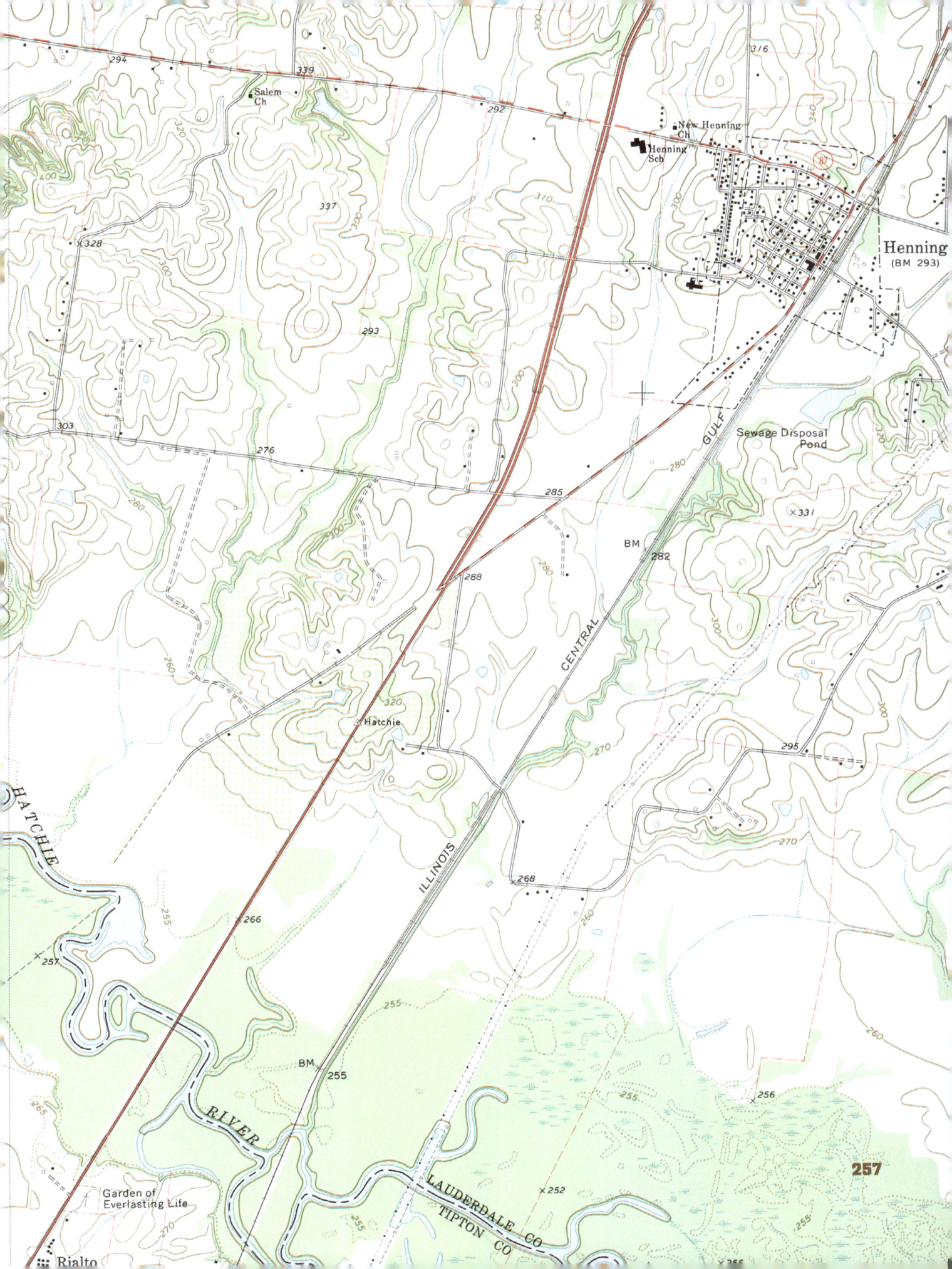

RIPLEY SOUTH, TENNESSEE

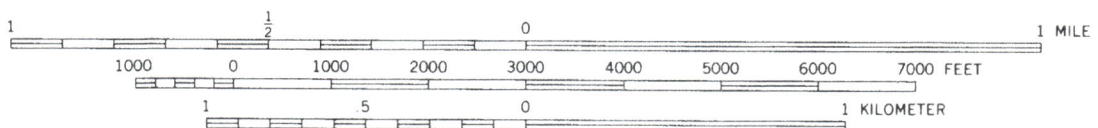

SCALE 1:24,000

Contour Interval 10 Feet

8. Approximately what portion of the floodplain will be submerged during the 50 year flood? (This can be determined by calculating the distance floodwater extends from the active channel to the edge of the floodplain.)

   _____

9. The Garden of Everlasting Life cemetery is located on the south side of the river. At what stage will flooding cover the cemetery?

   _____ feet

   Will water reach this cemetery during a 100-year flood? _____

10. Calculate the probability that a levee built to hold a 100 year flood will be over-topped by flood-water during

    any one year: _____

    a 10 year period: _____

    a 50 year period: _____

11. The size of flood control structures such as dams and levees are usually based on the 100 (or 500) year flood. Do you regard this as a reliable method to determine the height of flood-control levees? Explain.

A Natural Event

Historical data of a natural event will be provided by the lab instructor. The event is:
_____

Record the data on the following table and calculate the recurrence interval. After calculating the return interval, plot the data on Figure 22.4 and answer the questions that follow.

Year	Event	Rank	RI

Based on Figure 22.4 answer the following questions.

1. What is the magnitude of the predicted

    50 year event? _____

    100 year event? _____

    200 year event? _____

2. What is the probability that for any one year there will be at least one event no less (or greater) than:

    (magnitude) _____ : _____ %

    (magnitude) _____ : _____ %

    (magnitude) _____ : _____ %

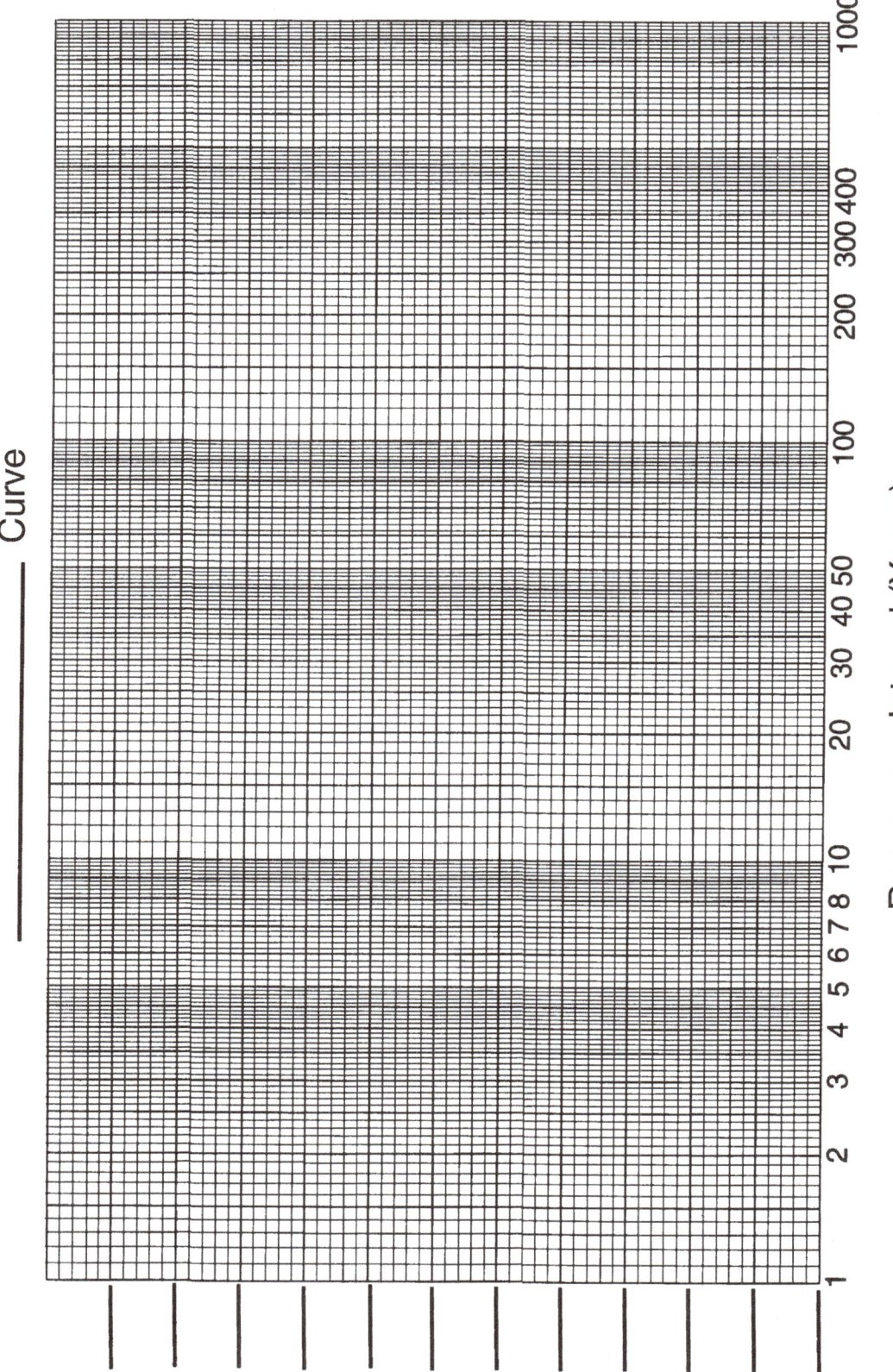

FIGURE 22.4 Historical data.

Name _____   Section _____

Fluvial Geomorphology

LAB EXERCISE 23

Running water is a dominant factor shaping the earth's surface. The nature of most landforms can be related to stream processes. Other means of erosion, such as glaciation, waves, and wind, may dominate in some areas, but none have as significant an impact over a large area of the earth. This section describes stream processes and erosional and depositional landforms created by running water.

Water flows downhill until reaching *base level*, generally at the elevation of the mouth of a stream where, where without velocity, there is no erosion. The ability of a stream to erode and transport sediment depends on discharge, which is the volume of water flowing past a given point in a specified period of time. Channel discharge is calculated by velocity multiplied by its cross-sectional area and is measured in cubic meters (or cubic feet) per second.

$$\frac{\text{Discharge}}{(m^3/\text{sec.})} = \frac{\text{velocity}}{(m/\text{sec.})} \times \frac{\text{cross-sectional area}}{(m^2)}$$

The velocity of water in a channel depends on both channel gradient and morphology that determines frictional resistance to flow. The flow is generally slower along rocky streams with irregular banks compared to straight streams with more symmetrical channel sides and bed.

In humid environments, discharge increases downstream because of additional water from tributaries. Rivers become progressively deeper and wider, but have a lower downstream gradient, which is shown below in the cross-sectional or longitudinal profile (Figure 23.1).

Sediment enters the channel from tributaries and through erosion of the channel bed and banks. If the streams gradient, and therefore its discharge is fast enough to transport the same amount of sediment that is entering the channel, the river is balanced, or in a state of equilibrium. Under these conditions, the river is not downcutting. This condition rarely holds for long periods of time because of seasonal variations in discharge. Rivers are continually downcutting or depositing sediment which builds up its channel bed. But over long periods of time rivers can maintain a balanced state.

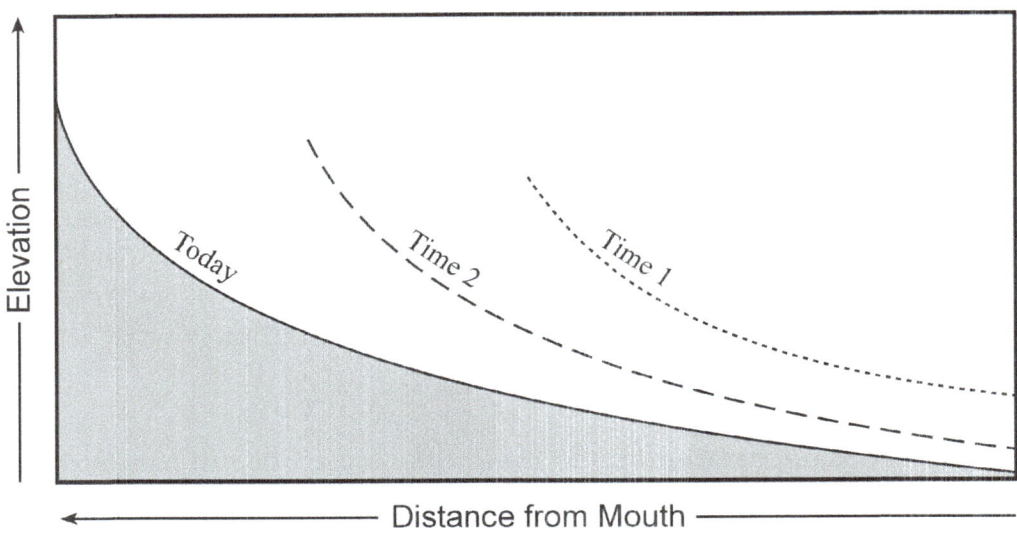

FIGURE 23.1 Longitudinal profile of a river.

1. **Stream Incision** (Reference: **Cape Solitude**, Arizona, Quadrangle)

   Streams with steep gradients have a high water velocity and the ability to erode or cut down the channel bed. These are referred to as **degrading streams**. The downward cutting creates steep V-shaped valleys with very little if any flat-land in the valley bottom. Rapids and waterfalls are common on these streams.

   The **Cape Solitude** map depicts the Colorado River where it passes through Marble Canyon upstream of the Grand Canyon. These streams have dissected the Colorado Plateau creating steep canyons with vertical cliffs and high local relief. Figure 23.2 is a stereogram of a portion of this map.

   a. What is the relief of the land area shown in the:

      map? _____

      photo? _____

   b. Cape Solitude is located south of the mouth of the Little Colorado River. What is Cape Solitude's:

      coordinates on the photo? _____

      elevation? _____

   c. How deep is the canyon below the rim at Cape Solitude? _____

   d. The mouth of Sixtymile Creek is about 2.5 km north of the mouth of the Little Colorado River. What are the coordinates on the photo for this position?

      _____

e. What is the stream gradient of Sixtymile Creek? _____

f. A delta has formed at the mouth of the Little Colorado River. What caused its formation?

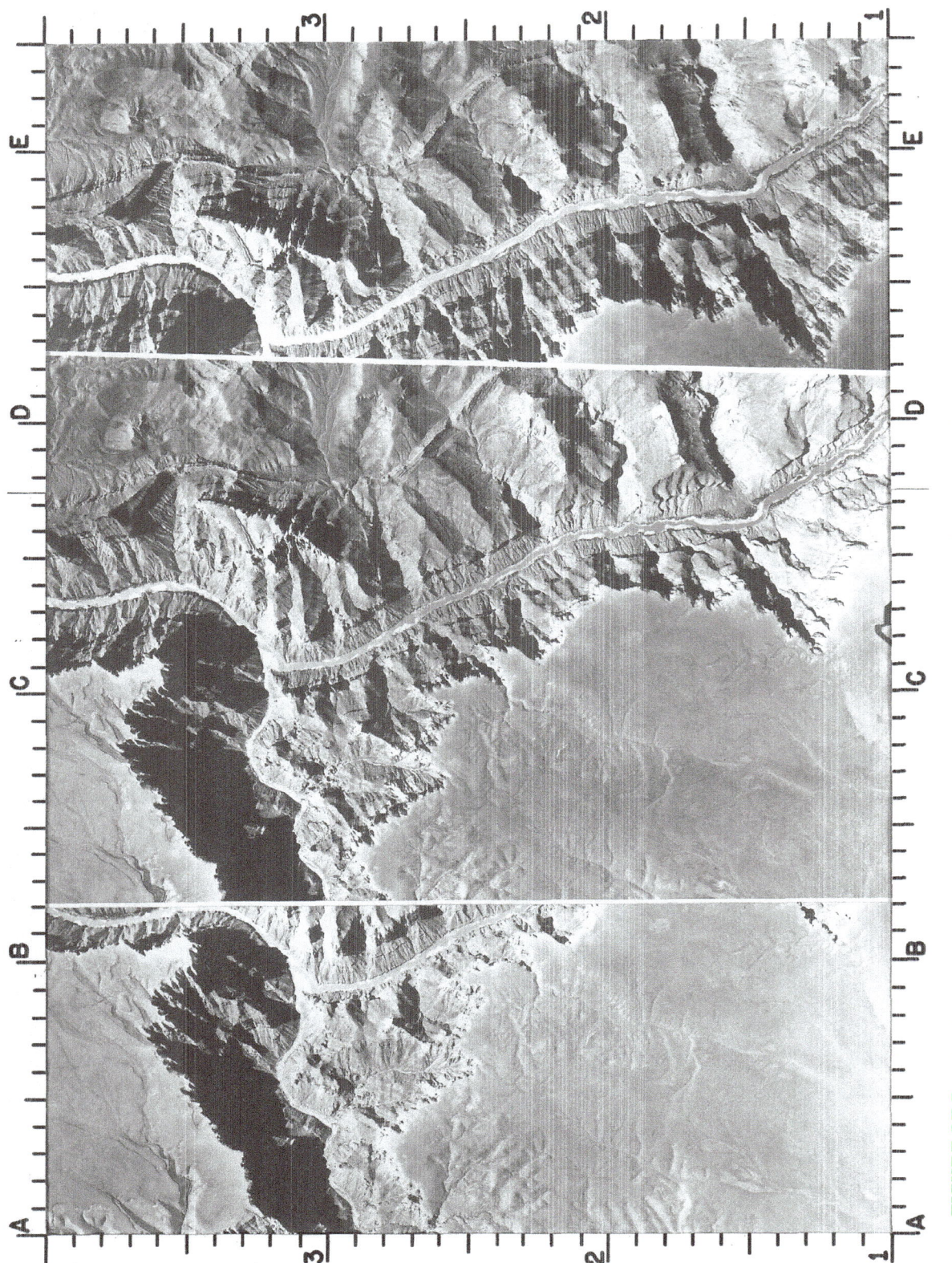

FIGURE 23.2 Marble Canyon in Colorado. Stereogram provided courtesy of the Map Library at the University of Illinois at Urbana-Champaign.

CAPE SOLITUDE, ARIZONA

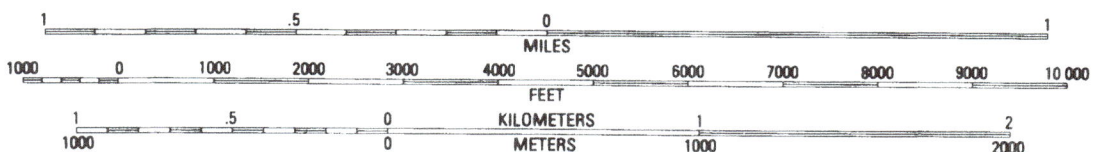

SCALE 1:24,000

Contour Interval 40 Feet

2. **Lateral Erosion** (Reference: **Walsh Knolls**, Utah-Colorado, Quadrangle)

As a stream erodes its channel bed, its elevation decreases. By cutting-down over long periods of time, the stream gradient becomes gentler and both water velocity and the rate of erosion slows. Eventually the elevation of the stream bed **stabilizes**. That is, the stream has reached a **state of equilibrium**, or a balance, between its ability to erode and the amount of sediment it carries. Once a stream is no longer downcutting it will begin **lateral erosion** and the creation of **meanders** that gradually widen the stream valley and create a **floodplain**.

The White River and many of its tributaries have dissected a plateau in western Colorado. The White River is now moving laterally and has created a narrow floodplain. Figure 23.3 is a stereogram of a portion of the **Walsh Knolls** map.

a. What is the relief of the land area shown in the:

map? _____

photo? _____

b. In which direction is the White River flowing? _____

c. Cottonwood Creek enters the White River in Section 24. What is the width of the White River floodplain at the mouth of Cottonwood Creek?

_____ m

d. The White River has begun to meander, indicating that it is no longer cutting down the channel bed. Are the tributaries of the White River down-cutting or moving laterally? Evidence?

e. The topographic map depicts roads and buildings. Are these or any other human features visible in the photo?

FIGURE 23.3 White River in Colorado. Stereogram provided courtesy of the Map Library at the University of Illinois at Urbana–Champaign.

WALSH KNOLLS, UTAH–COLORADO

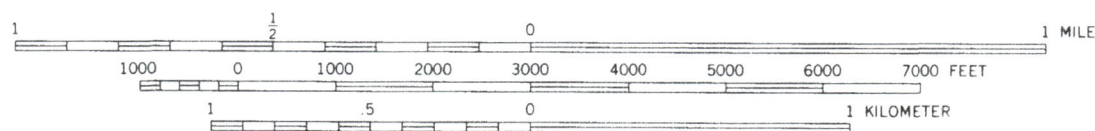

SCALE 1:24,000

Contour Interval 20 Feet

3. **Meandering Stream** (Reference: **Fishhook Lake**, Mississippi, Quadrangle)

   Meandering streams erode the surrounding valley sides and uplands, and over long periods of time can create broad, flat valleys. The flat surfaces adjacent to the meandering channel comprise its floodplain, and at least occasionally will be covered by floodwater. Almost all surfaces within these floodplains were at one time occupied by the active stream channel. As streams move laterally the meanders often come together creating a **cutoff**. The stream then changes course by flowing through the cutoff and leaving an **abandoned meander** or **oxbow lake**. Oxbow lakes eventually fill-in with sediment becoming **meander scars**. These former channels are depicted on topographic maps and in many cases distinguishable on aerial photographs.

   The **Fishhook Lake** map depicts the Tallahatchie River within a broad, nearly flat floodplain. Figure 23.4 is a stereogram of a small portion of this map. Many oxbow lakes and meander scars are visible on the photo.

   a. What is the relief of the land area shown in the:

      map? _____

      photo? _____

   b. The Tallahatchie River flows through central Mississippi, an area that normally supports forest vegetation. The Tallahatchie River floods almost every year, and yet most of the area in the floodplain is in cultivation. Why is there intensive agriculture in an area subject to flooding?

   c. Black Lake (A.6-1.5) is a former river channel. Why is the width of Black Lake in the photo smaller than that of the active channel?

   d. The aerial photographs (Figure 23.4) were taken in 1941. The corresponding topographic map was made from photographs taken in 1978, 37 years later. By comparing photographs and the map you can determine in many ways how the landscape was changed during this period. How have the following features changed?

      (1) active channel

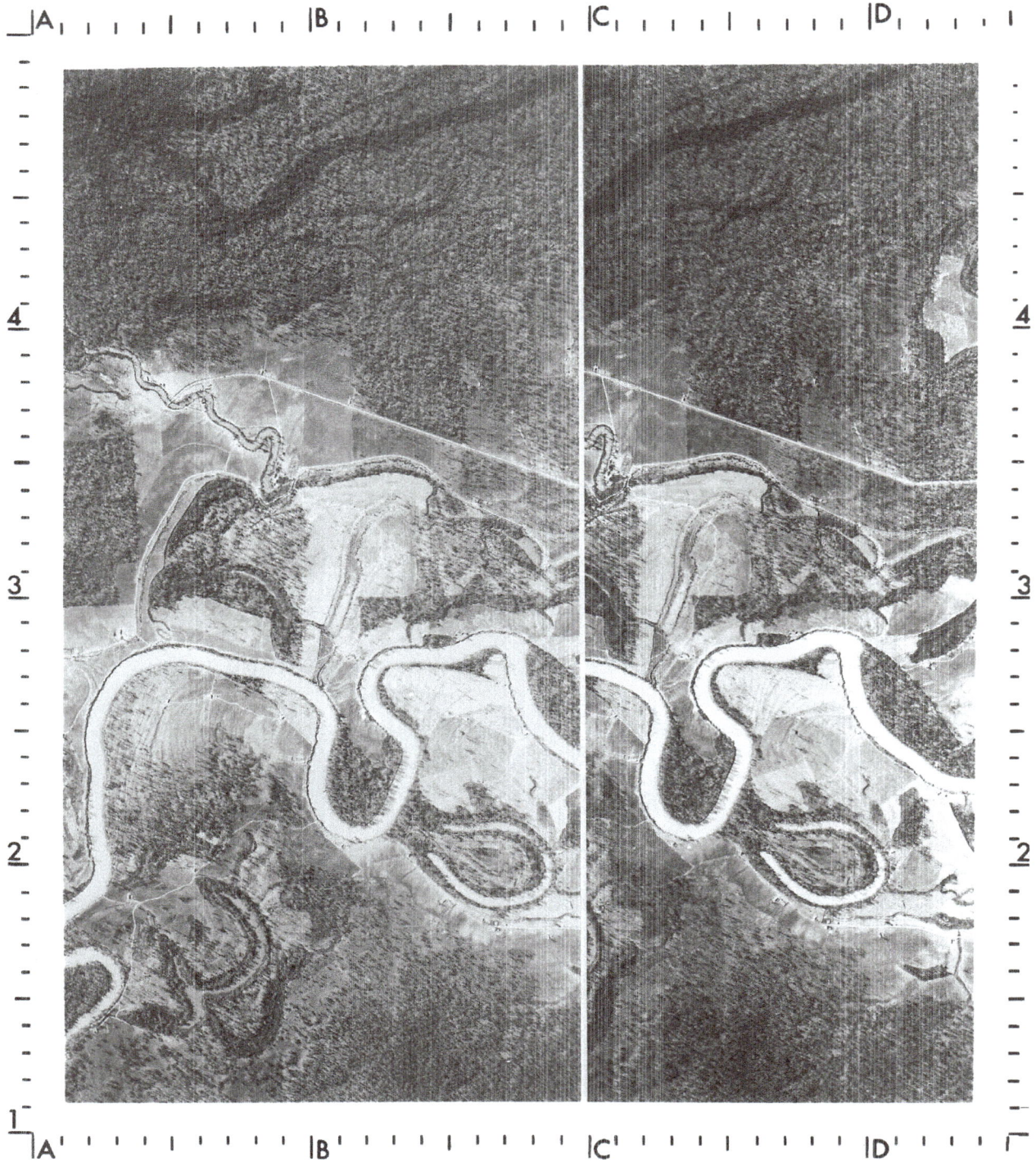

FIGURE 23.4 Horseshoe Lake in Mississippi. Stereogram provided courtesy of the Map Library at the University of Illinois at Urbana-Champaign.

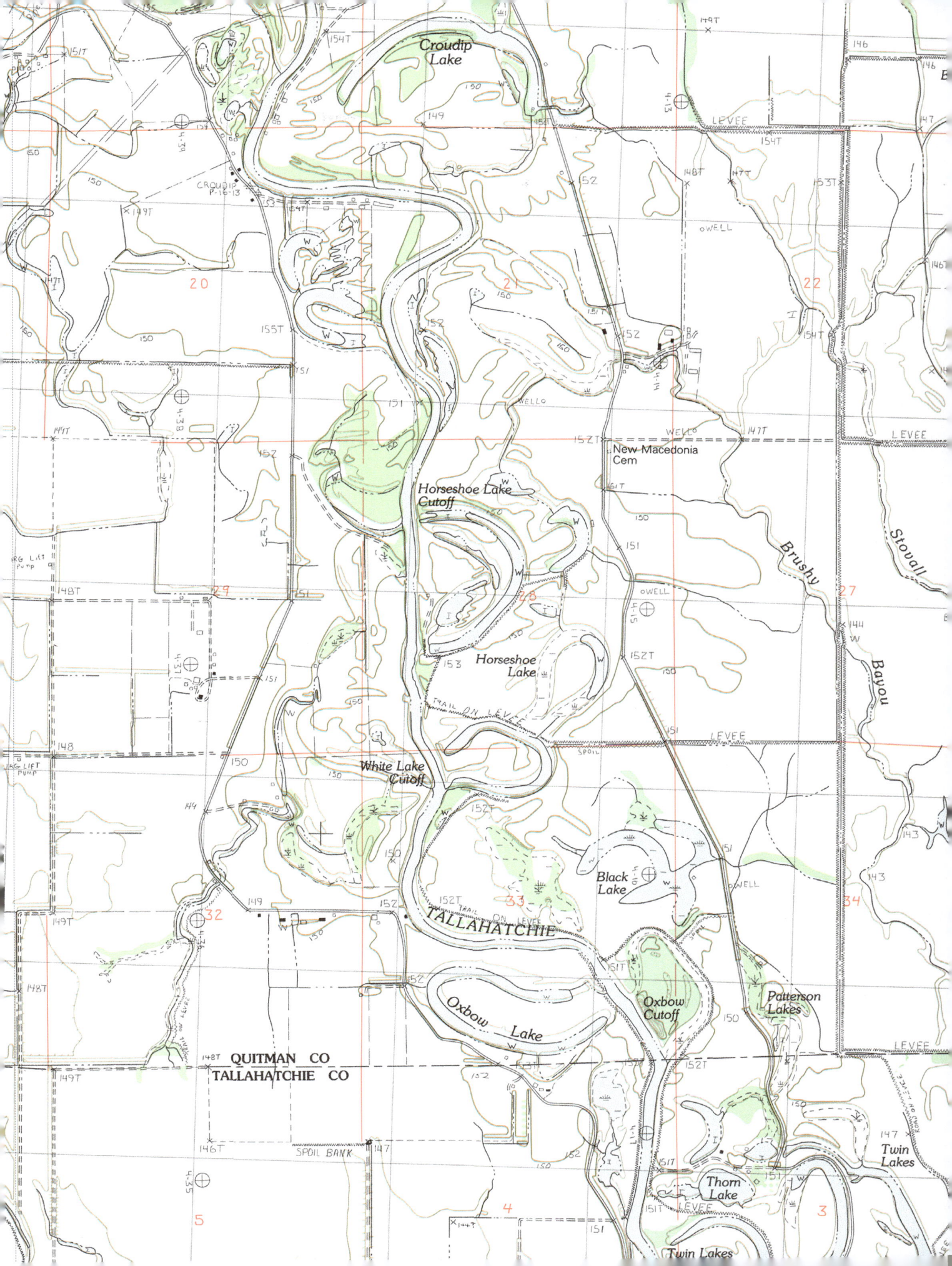

FISHHOOK LAKE, MISSISSIPPI

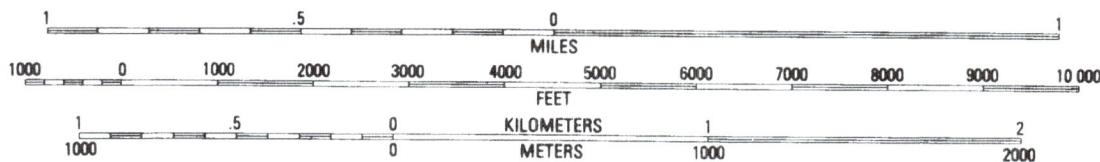

SCALE 1:24,000

Contour Interval 5 Feet

(2) Horseshoe Lake

(3) forest vegetation

4. **Stream Rejuvenation** (Reference: **The Loop**, Utah, Quadrangle)

    Streams that have created floodplains by lateral movement can renew downcutting if for any reason stream gradient increases. A steeper gradient will increase water velocity and result in a stream again eroding its channel bed. The renewed downcutting of a formerly meandering stream is defined as **rejuvenation**, which can be caused by (1) tectonic uplift, (2) decrease in base level, or (3) a change in climatic conditions that result in greater runoff. The stream will downcut to a lower elevation and may eventually create a new floodplain.

    The Loop map depicts the Colorado River flowing through a deep canyon. The former meandering stream began to downcut, resulting in **channel incision** and the **entrenchment** of meanders. Figure 23.5 is a stereogram of a small portion of this map.

    a. What is the relief of the land area shown in the:

    map? _____

    photo? _____

    b. In which direction is the Colorado River flowing? _____

    c. What is the width of the meander neck (land surface between the channels) at C.9-2.4?

    _____ m

    d. Why is no floodplain evident along this stretch of the river?

    e. Is there any evidence that the Colorado River tributaries have also entrenched? Explain.

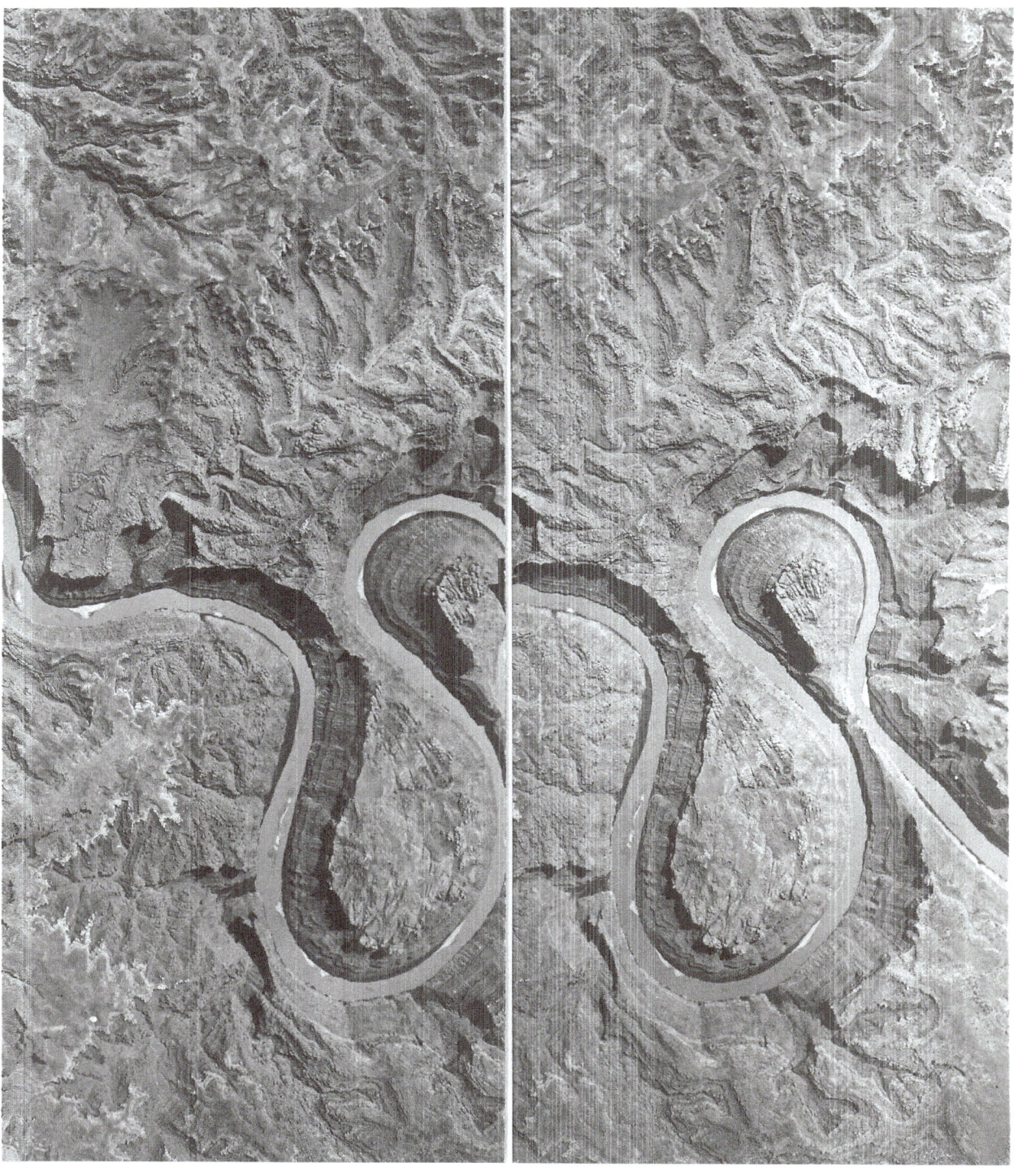

FIGURE 23.5 The Loop in Utah. Stereogram provided courtesy of the Map Library at the University of Illinois at Urbana-Champaign.

THE LOOP, UTAH

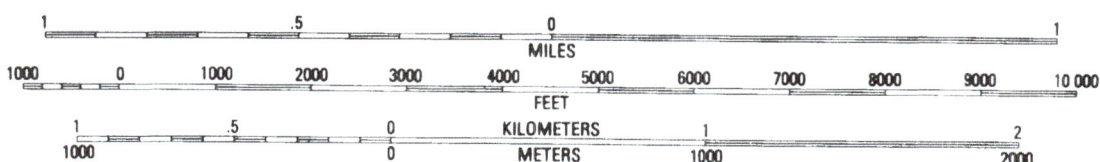

SCALE 1:24,000

Contour Interval 40 Feet

5. **Braided Stream** (Reference: **Nevens**, Nebraska, Quadrangle)

   In contrast to meandering streams, **braided streams** have multiple, shifting channels. Low-gradient streams typically develop this type of river pattern if they have very high sediment loads. Channels are separated by sand or gravel bars. High discharge will redistribute the sediment causing laterally shifting channels.

   The Nevens map depicts a braided section of the North Platte River in Nebraska. This stream is flowing toward the Mississippi River to the east. It is fed by snow melt-water from the Rocky Mountains. Figure 23.6 is a stereogram of a small portion of this map.

   a. Vegetation on the mid-channel bars separating channels would suggest the bars are somewhat stable. Can you distinguish if bars are vegetated or not? Explain.

   b. There is a single water channel in the right photograph at the edge of the North Platte River valley. Is this a natural channel or man-made? Explain.

   c. There is vegetation within the river banks. But as shown on the right photograph, there is little vegetation on the surrounding upland. Explain.

   d. Is there evidence of agriculture on the photograph?

FIGURE 23.6 North Platte River in Nebraska. Stereogram provided courtesy of the Map Library at the University of Illinois at Urbana-Champaign.

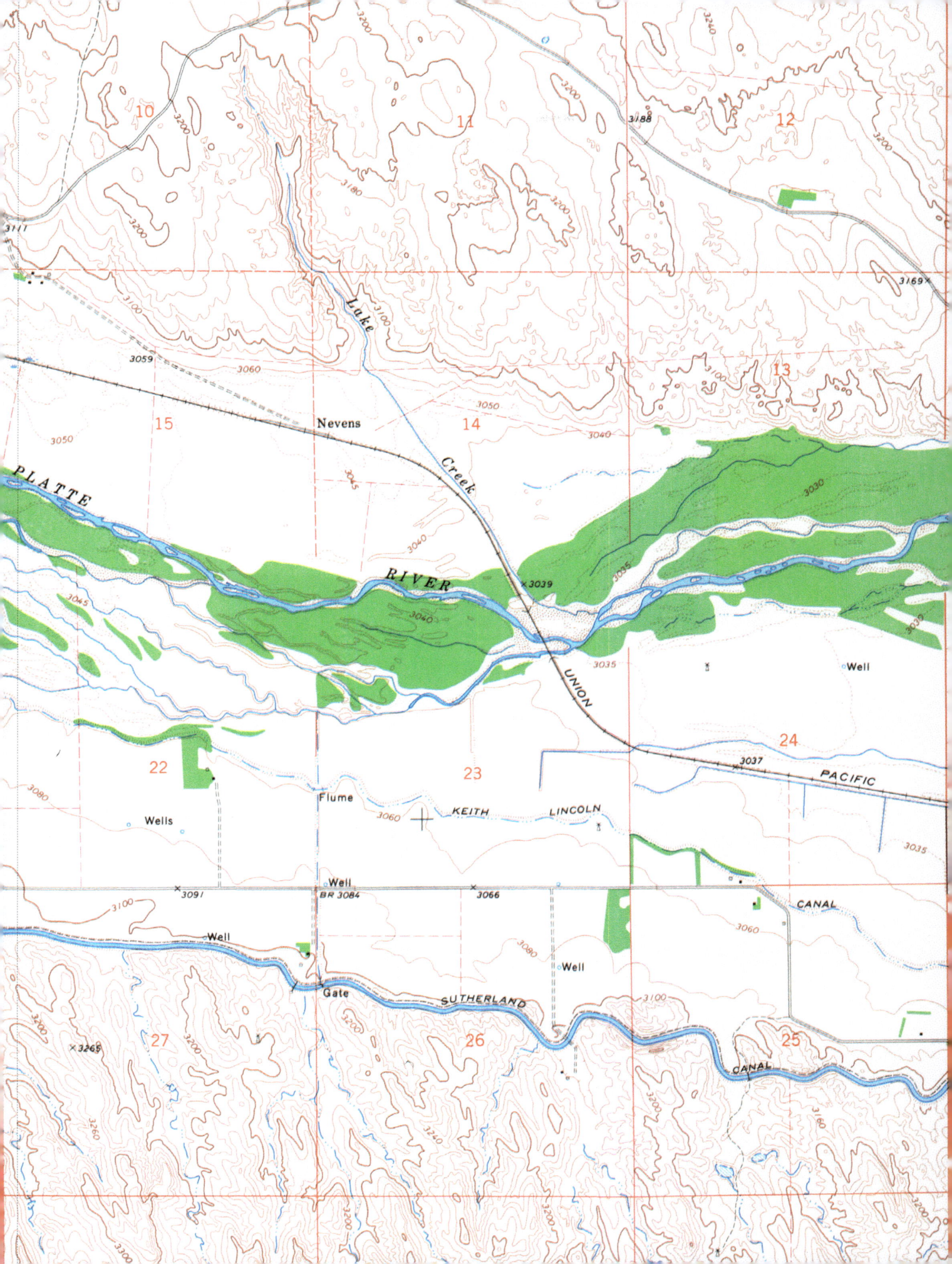

NEVENS, NEBRASKA

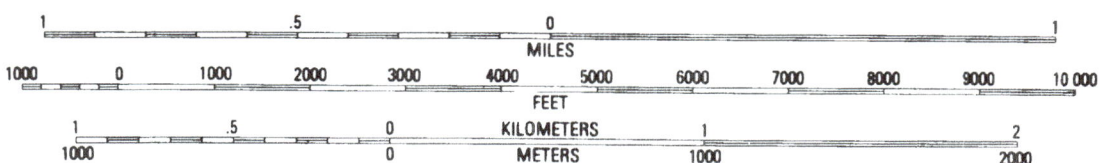

SCALE 1:24,000

Contour Interval 20 Feet

Name _____ Section _____

LAB EXERCISE 24

Arid Landscapes

Arid Landscapes

Dry regions of the world have distinctive physical features. In many places soils are either thin or absent and vegetation is sparse. Most streams are ephemeral, flowing only during or after rainfall or from melting snow in surrounding mountains. No permanent streams originate in desert or semi-desert regions. Also, deserts contain rivers that flow into basins or valleys and do not drain into any ocean. In deserts, evaporation exceeds precipitation and rivers become progressively smaller with distance downstream.

Although most desert streams are ephemeral, water is the dominant agent shaping the landscape. Vegetation provides little cover to intercept rain and the absence of surface organic matter and root systems allows erosion and rapid movement of sediment. This section describes landforms in dry climates that include desert and steppe regions.

1. **Alluvial Fans** (Reference: **Stovepipe Wells**, California, Quadrangle)

   Some streams have such a high sediment load that not all the material can be transported downstream. Instead the sediment is deposited which raises the channel bed elevation. The **alluvial fans** depicted on the **Stovepipe Wells** map have formed at the edge of Death Valley. They develop when ephemeral streams flowing in the surrounding mountain canyons emerge into the desert basin. Because of the lower stream gradient in the desert basin, water velocity slows and the sediment is deposited. Figure 24.1 is a stereogram of a small portion of this map.

   a. What is the stream gradient in Mosaic Canyon? _____

   b. What is the gradient of the same stream after it flows out of the mouth of the Canyon (D.1-2.5)?
   _____

FIGURE 24.1 Stovepipe Wells in California. Stereogram provided courtesy of the Map Library at the University of Illinois at Urbana-Champaign.

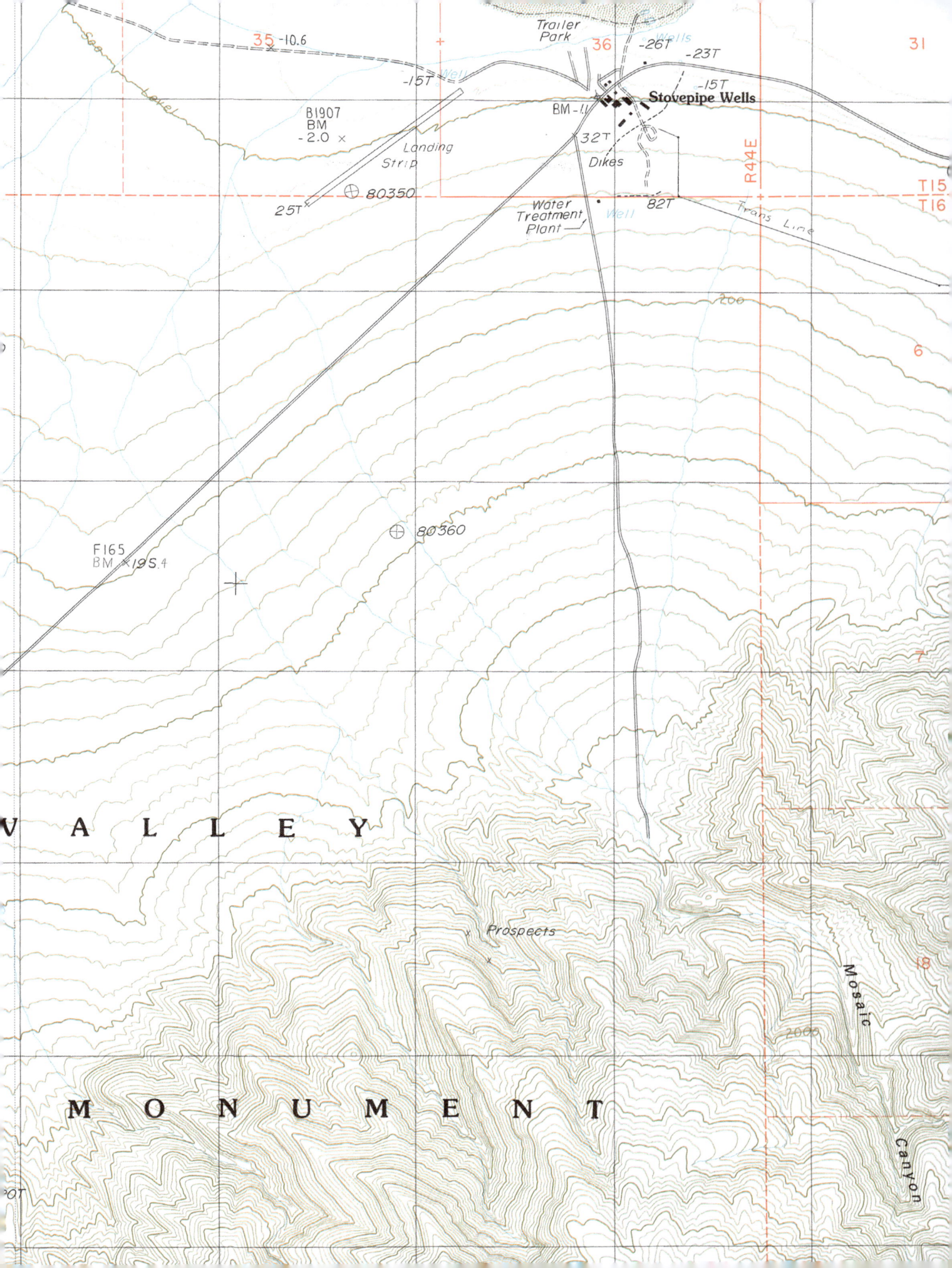

STOVEPIPE WELLS, CALIFORNIA

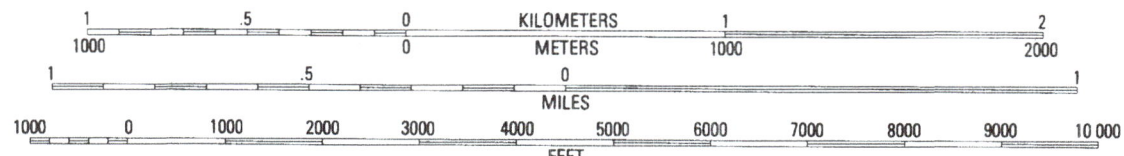

SCALE 1:24,000

Contour Interval 40 Feet

c. Multiple stream channels on the alluvial fans can easily be seen on the photo. Why are there so many channels on these surfaces as opposed to only one main channel in the canyon?

d. What is the dark line near the edge of the fans (B.9-3.0)?

e. What is the nature of vegetative cover on and beyond the edge of the fans?

f. What are the photo coordinates for the town of Stovepipe Wells?

   _____

2. **Meandering Stream and Headward Erosion** (Reference: **Scenic**, South Dakota, Quadrangle)

   As mentioned earlier, dry regions have sparse or no vegetation. Although rain is infrequent, erosion can be severe compared to humid regions with more dense vegetation. To contrast erosion potential in humid and dry regions, consider how forest vegetation modifies hydrology. First, the forest canopy intercepts rain. During heavy rain water will eventually begin dripping through the canopy. The surface organic layer and the upper mineral soil horizon that is mixed with organic material effectively holds water. Only after long periods of rain do the upper soils become saturated. If surface infiltration is no longer possible, water begins to move across the ground surface, potentially moving sediment. In most dry regions, water will move across the surface much more quickly, which has notable effect on the landscape, as shown in the following photographs.

   The **Scenic** map depicts a small river with a tortuous meandering pattern in South Dakota. Tributaries are extending their valleys into the surrounding uplands. Figure 24.2 is a stereogram of a small portion of this map.

   a. Has this river developed a floodplain? Evidence?

   b. What is the ratio of the width of the floodplain to the width to the meander belt?

FIGURE 24.2 Bear Creek in South Dakota. Stereogram provided courtesy of the Map Library at the University of Illinois at Urbana-Champaign.

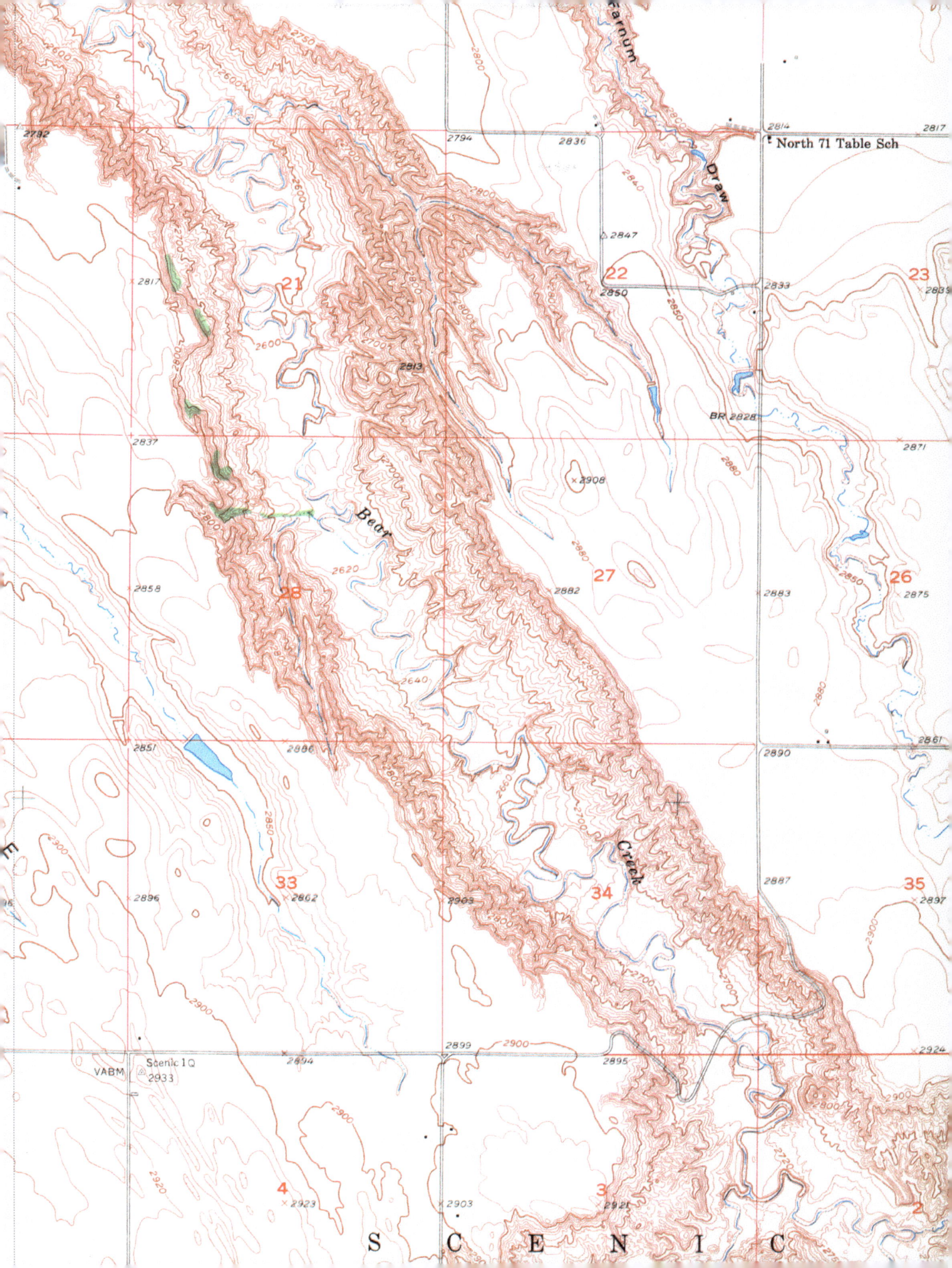

SCENIC, SOUTH DAKOTA

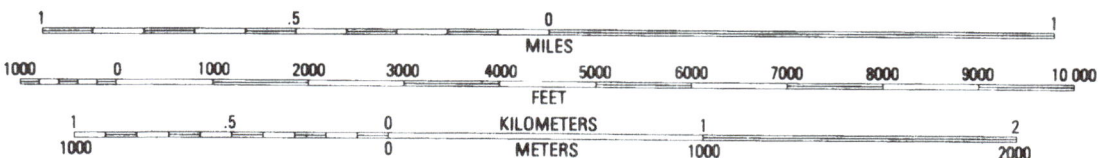

SCALE 1:24,000

Contour Interval 10 Feet

c. Have the smaller tributary streams developed floodplains? _____

d. What is the difference in elevation between the floodplain and surrounding plateau?
_____

e. Is there any agriculture in the area? Evidence?

f. Are any trees shown in the photograph? Where?

3. **Dissection of a Plateau** (Reference: **Sallies Spring**, New Mexico, Quadrangle)

   As shown on the previous map and photograph, the **Sallies Spring** map depicts tributaries of a small meandering river that are incising and extending their valleys into the surrounding uplands. In this case, overlying rock of the upland plateau is highly resistant to erosion resulting in the development of cliffs. Note the shallow stream pattern at the top of the photograph. This stream is at a much higher elevation than Bear Creek, but water drains away from the river. Figure 24.3 is a stereogram of a small portion of this map.

   a. What is the difference in elevation between the San Juan River floodplain and plateau near the Drill Hole south of the river?
   _____

   b. What is the width of the

   meander belt of San Juan River? _____ m

   floodplain of the San Juan River? _____ m

   c. Is there any evidence of sediment deposition by the San Juan River? What depositional features are present?

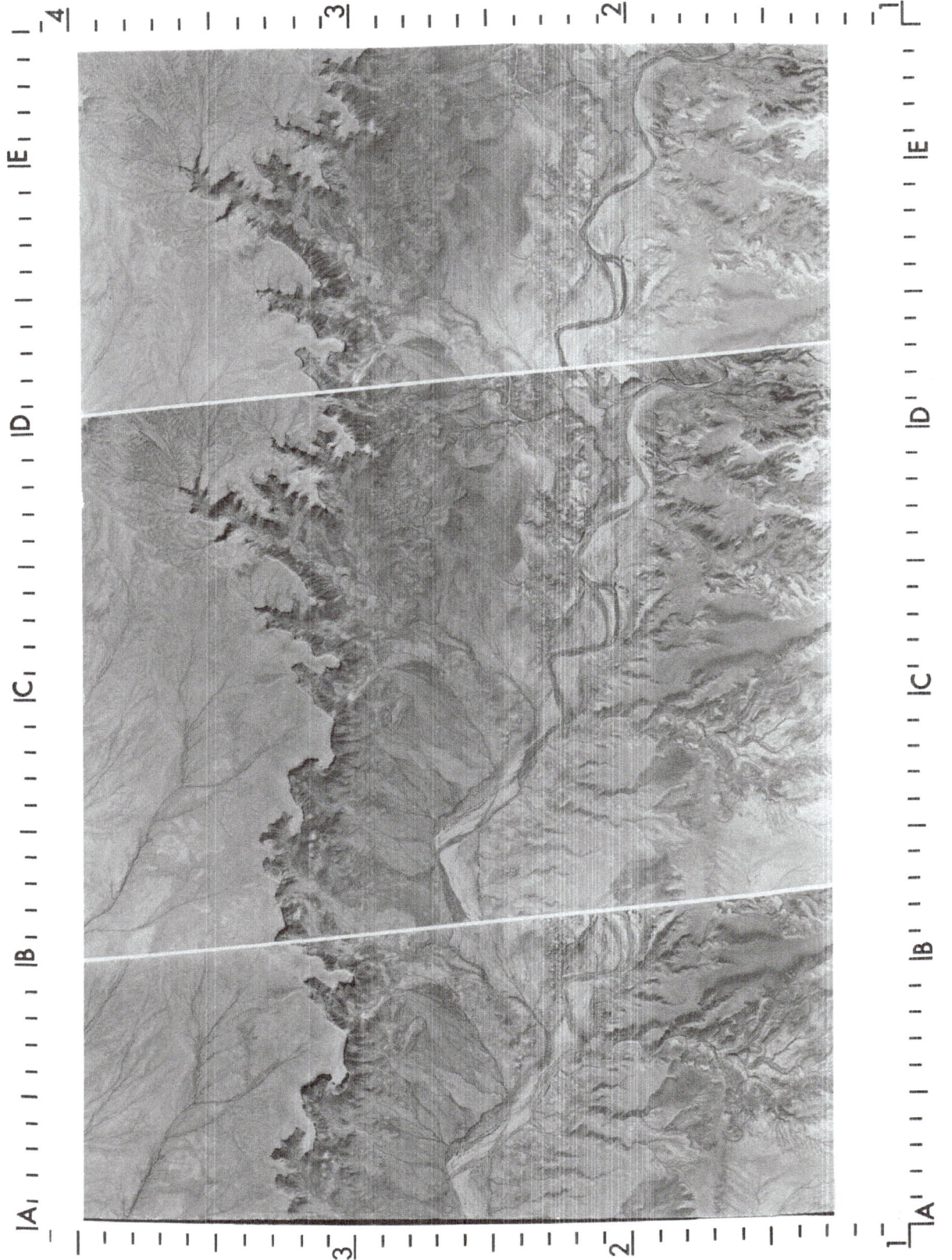

FIGURE 24.3 San Juan River in New Mexico. Stereogram provided courtesy of the Map Library at the University of Illinois at Urbana-Champaign.

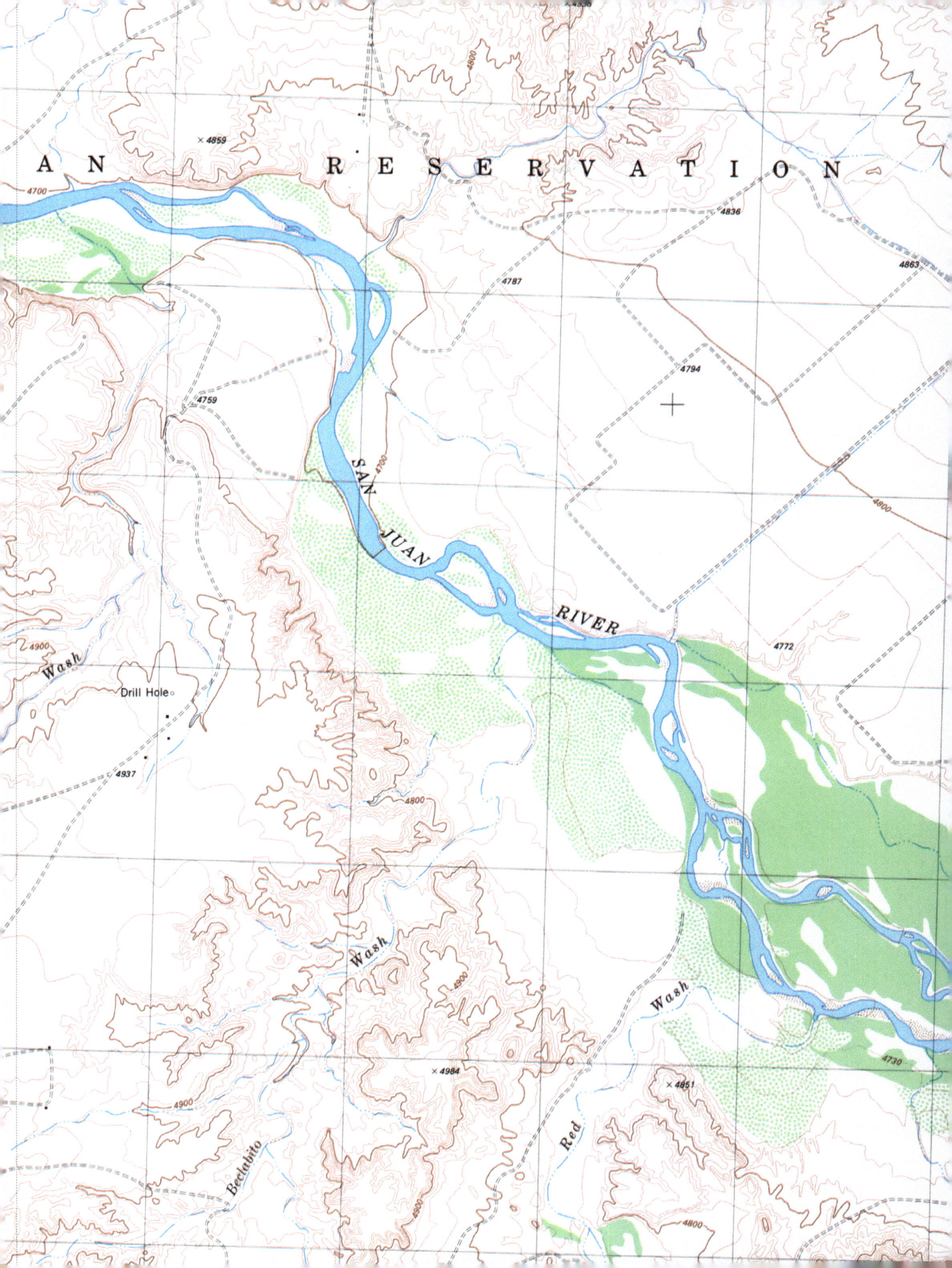

SALLIES SPRING, NEW MEXICO

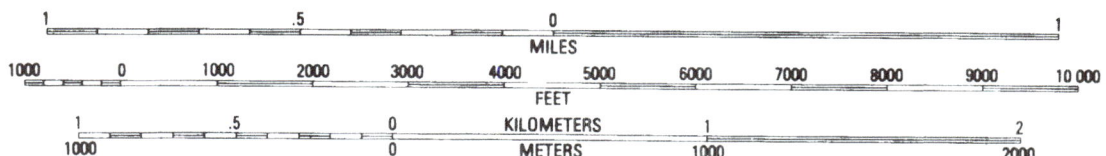

SCALE 1:24,000

Contour Interval 20 Feet

d. The photographs were taken in September, one of the driest months of the year in New Mexico. At the time of the photograph was water flowing through the:

San Juan River? _____

tributaries of the San Juan River? _____

e. What is the gradient of tributary streams before they enter the San Juan River floodplain?

4. **Remnant Mountains** (Reference: **Wister**, California, Quadrangle)

True desert regions have almost no surface water and no permanent streams, except for those that originate in wetter climates and then flow into the desert. Regardless, water is still important in shaping the landscape and stream courses are easily observed because of the sparse vegetation. An area of southeastern California is depicted on the **Wister** map. This region is in the rainshadow of the Sierra Nevada Mountains and is one of the driest areas in North America. Figure 24.4 is a stereogram representing a section of the **Wister** map showing highly dissected mountains surrounded by eroded debris. There are many temporary stream courses moving weathered rock debris that has covered all but the tops of these mountains.

a. Which direction is the debris moving?

b. Describe the vegetation of this region.

c. Is there evidence of:

permanent streams? _____

agriculture? _____

human occupation? _____

d. Roads are shown on the topographic map. Are they visible in the photographs? _____

e. Note the man-made canal on the lower section of the map. Few people live in this region so what is the likely purpose of the canal?

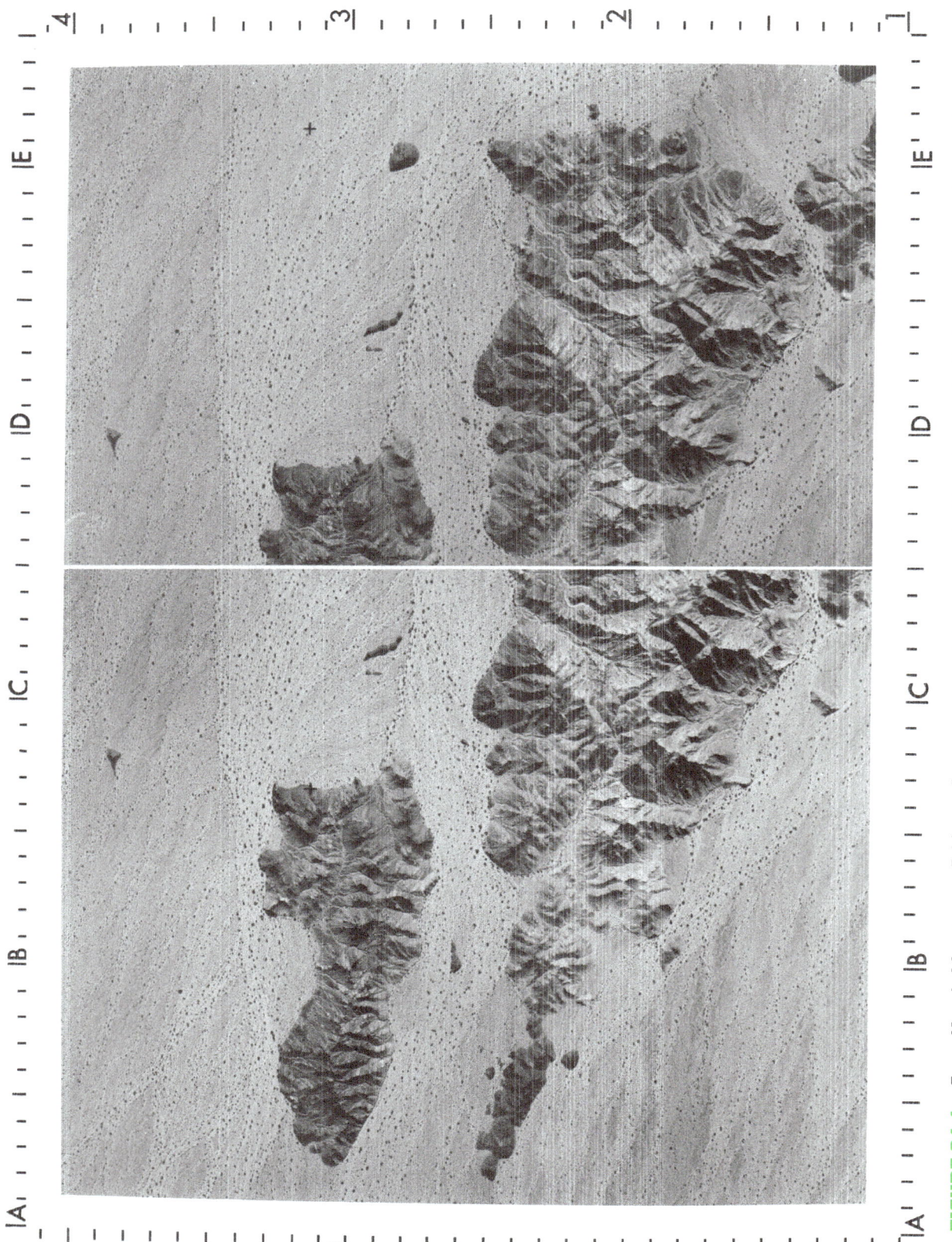

FIGURE 24.4 Sierra Nevada Mountains in California. Stereogram provided courtesy of the Map Library at the University of Illinois at Urbana-Champaign.

WISTER, CALIFORNIA

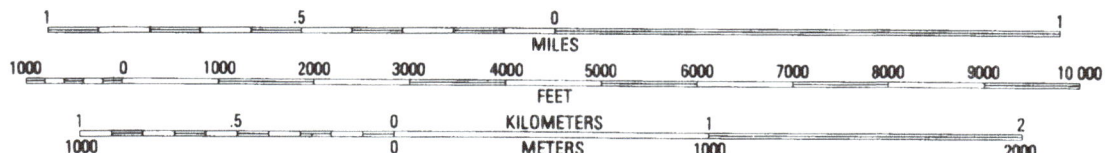

SCALE 1:24,000

Contour Interval 10 Feet

Name_____ Section_____

LAB EXERCISE 25

Glaciated Landscapes

Glaciated Landscapes

Much of the land surface of the mid-high latitudes in both the northern and southern hemispheres have been affected by glacial processes. **Continental glaciers** with thick ice sheets advanced across much of northern Europe, Asia, and North America several times during cold periods of earth's history. In eastern North America, glaciers extended as far south as the Ohio River valley. High mountain regions throughout the world were occupied by **mountain glaciers** that moved down steep valleys. The scouring of the earth's surface by ice and deposition of glacial sediment created distinct topographic features.

The past 2 million years has been a cold period. This most recent geologic epoch known as the **Pleistocene** ended about 10,000 years ago. During this time there were successive glacial periods interrupted by warm periods when the ice retreated.

Now, glaciers cover only a small percent of the earth's surface compared to most of the Pleistocene. However, most of Antarctica, Greenland, in addition to a few other areas in the high latitudes of the northern hemisphere are still covered by ice. Also, mountain glaciers commonly occur in most of the very high mountain ranges throughout the world. So glacial processes are still shaping the landscape.

1. **Mountain Glacier** (Reference: **Mt. Fairweather**, Alaska, Quadrangle)

   Snow accumulation in high mountain regions will compact into ice, and if thick enough, it will begin to move down the valleys. The ice scours the valley as it moves downslope and therefore, alpine glaciers typically carry large amounts of sediment. Glaciers melt as they move to lower elevations. If the rate of snow accumulation at the source of the glacier high in the mountains exceeds the rate of melting as the glacier descends, the glacier front will advance downslope. Conversely, if the rate of melting exceeds snow accumulation, the glacial front will retreat up the valley.

   Crillon Glacier in Alaska is depicted on the **Mt. Fairweather** map. **Note:** The scale for this map is 1:250,000. The scale for the other topographic maps is 1:24,000. Figure 25.1 is a stereogram showing the lower section of this glacier.

FIGURE 25.1 Crillon Glacier in Alaska. Stereogram provided courtesy of the Map Library at the University of Illinois at Urbana-Champaign.

MT. FAIRWEATHER, ALASKA

Contour Interval 250 Feet

a. What is the elevation of Mt. Crillon? _____

b. What causes the dark bands on the glacier? (Hint: The dark band that extends from the point where the north and south glaciers merge.)

c. A delta has developed at the end of the glacier. Explain the development of this delta.

d. The end of a smaller glacier in a higher valley can be seen on the photograph (C.3-2.5). Is there evidence of melt-water flowing from this glacier? Explain.

e. Ice has plastic characteristics and the glacier can change shape as it flows downslope. As it moves, the surface of the glacier will fracture causing crevasses to develop. Give the photograph coordinates for crevasses present on the Crillon Glacier.

_____

2. **Glacial U-Shaped Valley** (Reference: **Mt. Moran**, Wyoming, Quadrangle)

Erosion by mountain glaciers deepen and widen the valley floor. Streams will erode valleys with a cross section that is V-shaped. In contrast, mountain glaciers create U-shaped valleys. After the glacier retreats, streams will reoccupy the valley which is sometimes referred to as a **glacial trough**.

A U-shaped valley once occupied by a glacier in the Grand Teton National Park is depicted on the **Mt. Moran** map. Figure 25.2 is a stereogram showing the valley.

Note: Getting a clear stereo image of this valley is difficult. Try looking first at the lower part of the photographs. Be patient and you will get a very good view of the valley.

a. Give the elevation for:

Cascade Creek: _____

Lake of the Crags: _____

Mt. St. John: _____

FIGURE 25.2 Grand Teton National Park in Wyoming. Stereogram provided courtesy of the Map Library at the University of Illinois at Urbana-Champaign.

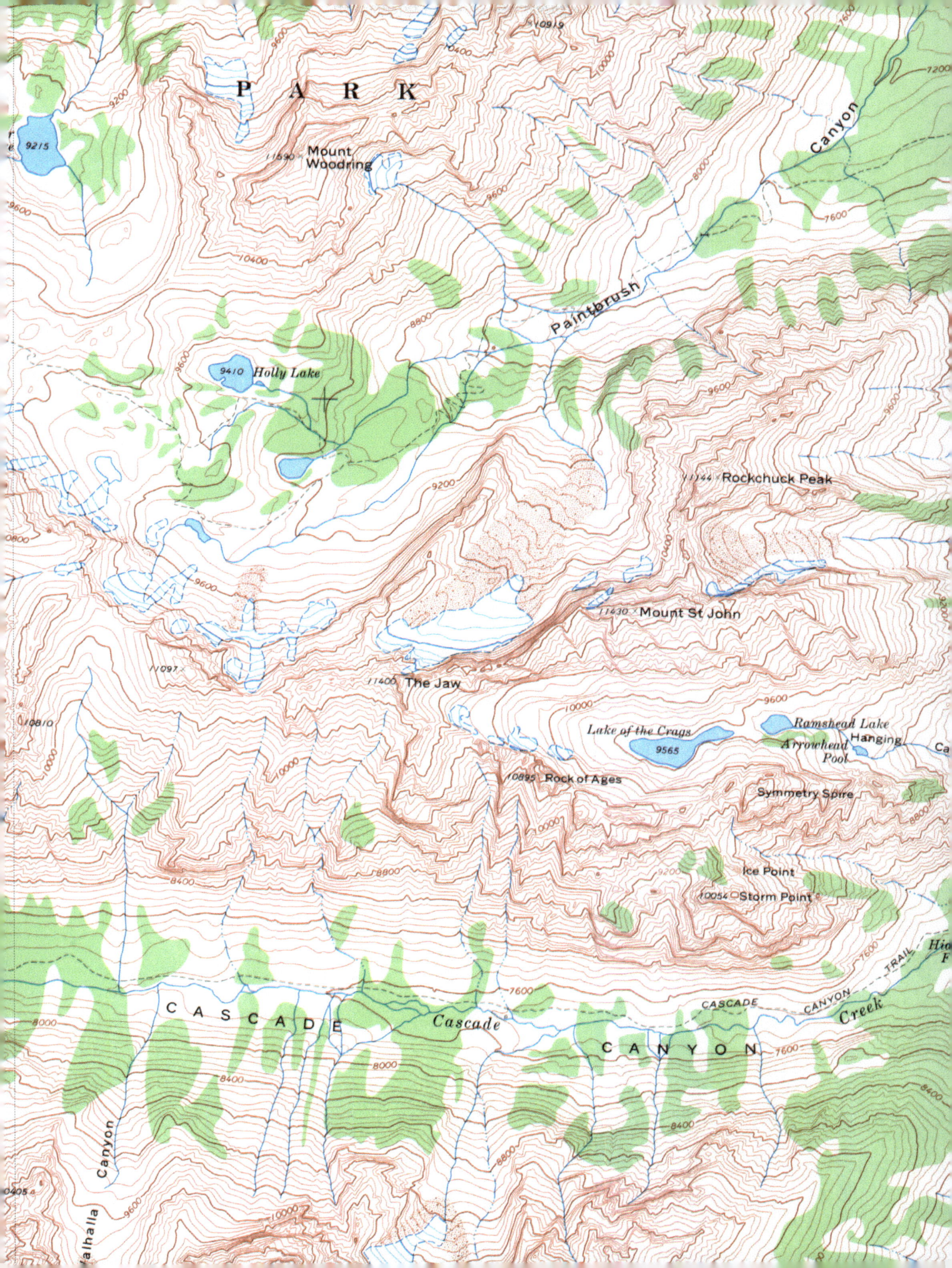

MT. MORAN, WYOMING

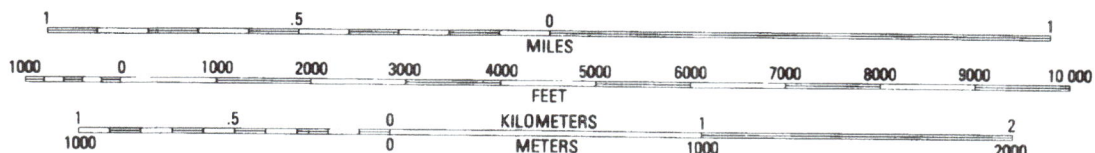

SCALE 1:24,000

Contour Interval 80 Feet

b. Is there snow in the photograph? Where?

c. Is snow melt-water flowing into the valley? Describe these streams.

d. Where does vegetation occur?

e. How has debris flow (landslides) affected the bottom of the valley?

3. **Surface Erosion by a Continental Glacier** (Reference: **Daley Bay**, Minnesota, Quadrangle)

   During the Pleistocene, thick continental ice sheets moved southward across Canada and the bordering area of the U.S. near the Great Lakes, including Minnesota, Wisconsin, Michigan, and New York. The glaciers scoured the landscape removing most soil and quarrying out surface rock. This process disrupted drainage patterns and left depressions. As a result, there are numerous lakes over a large area of central North America.

   The **Daley Bay** map depicts an area in Minnesota with many lakes occupying depressions caused by glaciation. Figure 25.3 is a stereogram of a small portion of this map.

   a. Grooves in rock made by the last glacier can be seen on the photograph near A.8-3.6. This indicates a movement of ice from either the left or right. Based on these grooves, which direction did the last glacier move? (Hint: Do not assume the top of the photograph is north. Refer to the topographic map to determine direction.)

   _____

   b. What is the total relief of the area depicted in the topographic map?

   _____

FIGURE 25.3 Lake Kabetogama in Minnesota. Stereogram provided courtesy of the Map Library at the University of Illinois at Urbana-Champaign.

DALEY BAY, MINNESOTA

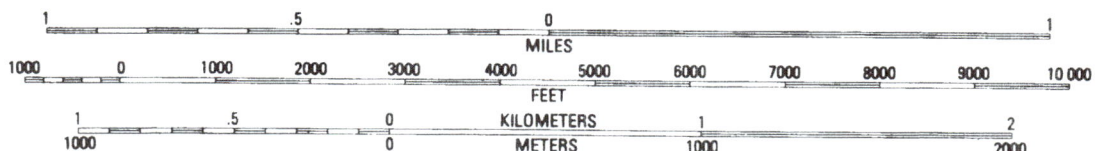

SCALE 1:24,000

Contour Interval 10 Feet

c. Is there evidence of human occupation? Explain.

d. Do you expect this area to support intensive agriculture? Explain.

4. **Drumlins** (Reference: **St. Peter**, Wisconsin, Quadrangle)

   Drumlins are small, elongated hills composed of unsorted glacial till. The long axis of the drumlins are parallel to the direction of glacial movement. Their form offers minimum resistance to the movement of glacial ice. But the exact formation processes are unknown.
   The **St. Peters Bay** map depicts an area in Wisconsin with many drumlins. Figure 25.4 is a stereogram of a small portion of this map.

   a. Based on the axis of the drumlins, in what direction was the glacier moving?

      _____

   b. What is the elevation of the summit of these drumlins?

      average: _____

      minimum: _____

      maximum: _____

   c. Is there evidence that at the time these photographs were taken that these drumlins are severely eroding? Explain.

   d. How has this landscape been modified by human activity?

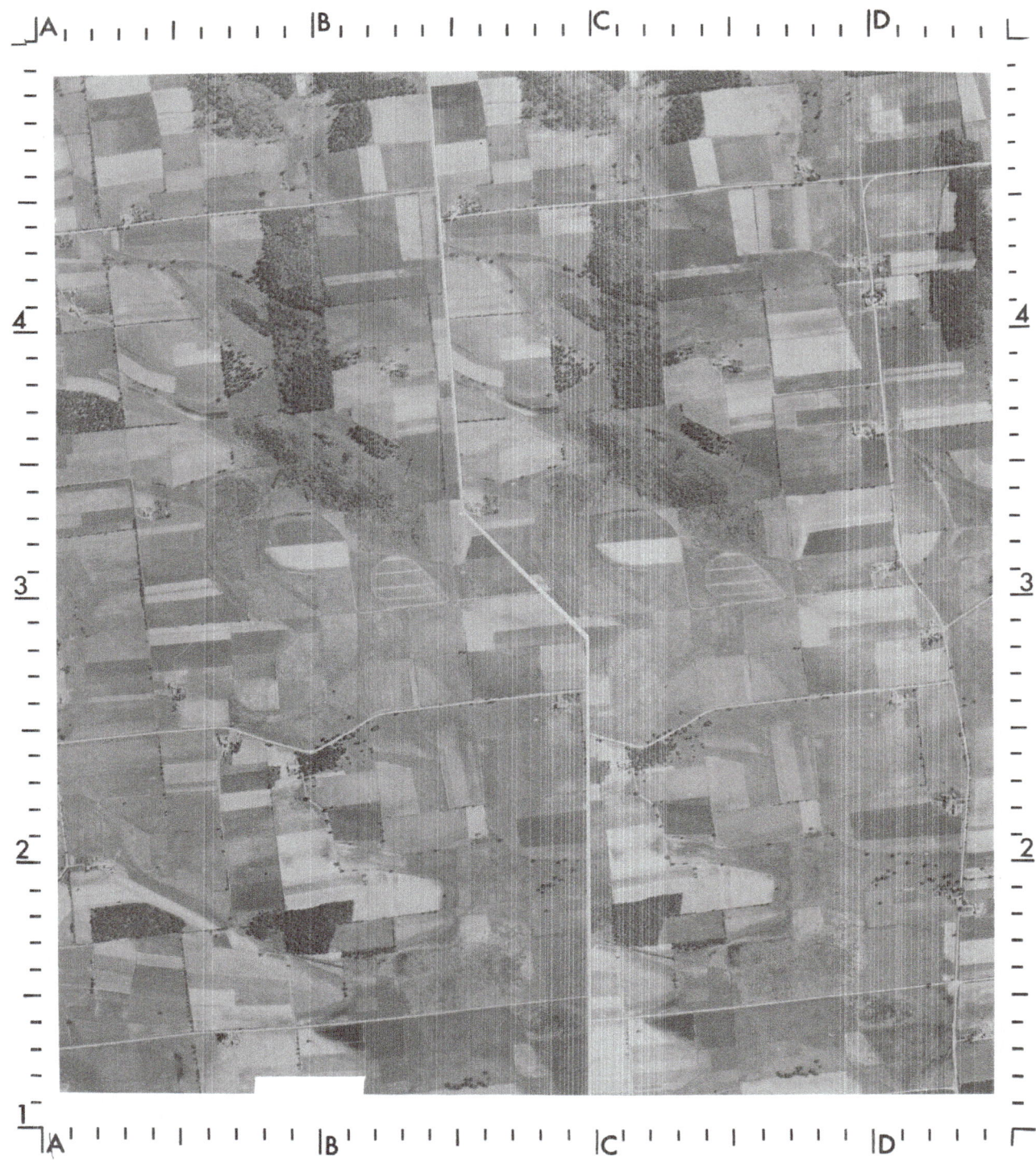

FIGURE 25.4 Drumlins in Wisconsin. Stereogram provided courtesy of the Map Library at the University of Illinois at Urbana-Champaign.

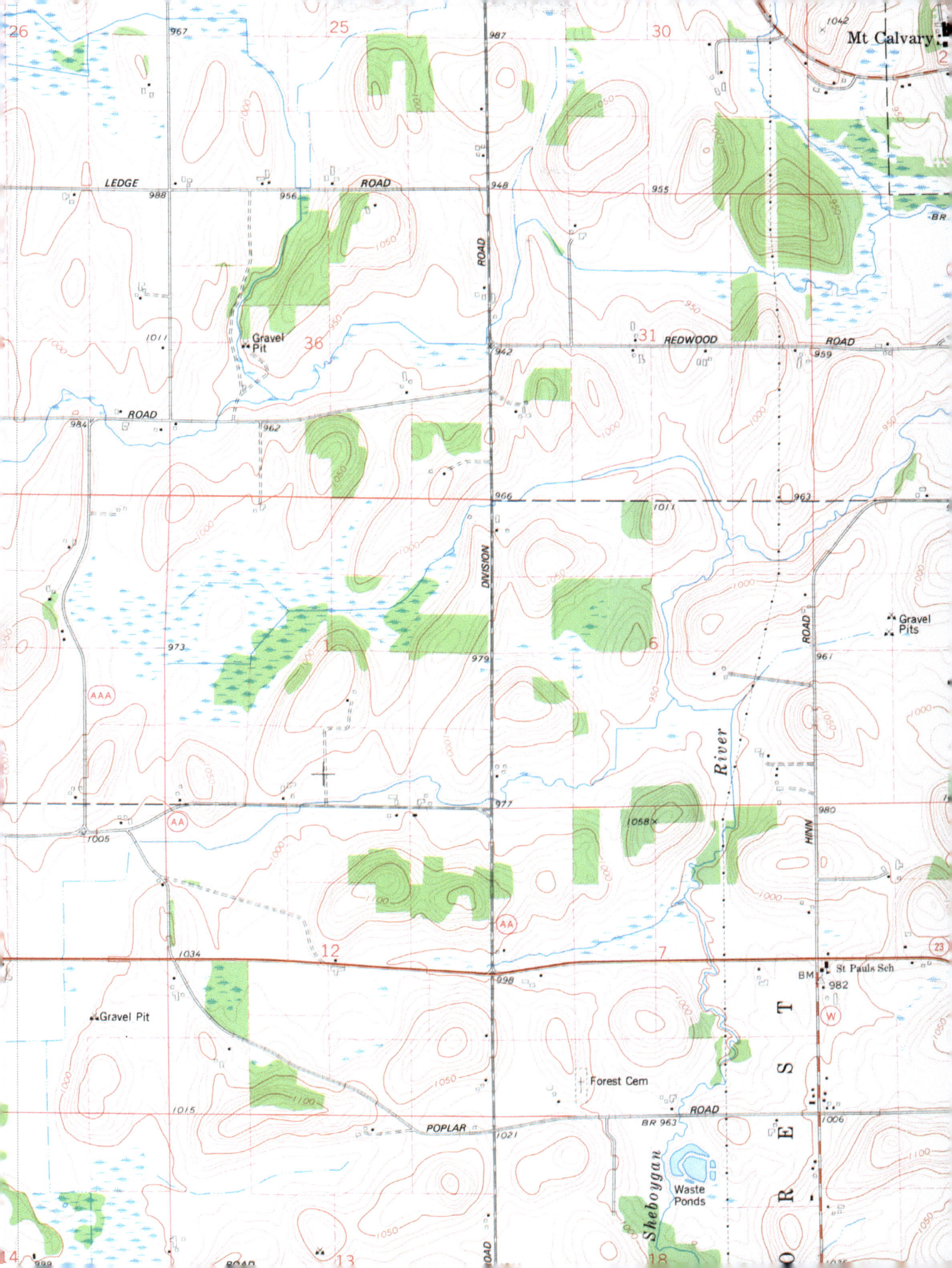

ST. PETERS BAY, WISCONSIN

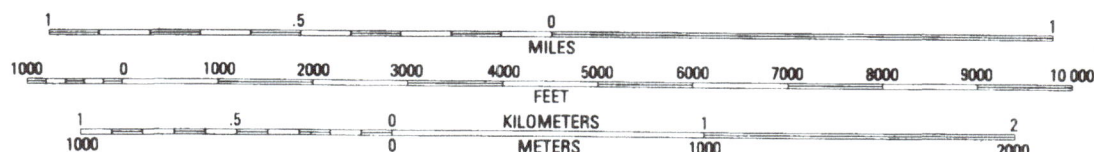

SCALE 1:24,000

Contour Interval 10 Feet

Name_____   Section_____

LAB EXERCISE
26

Coastal Geomorphology

Coastal Geomorphology

Ocean and lake coastlines extend for thousands of kilometers on each of the continents. Wave motion is the dominate force shaping these landforms, but other processes affect the coastal margins, including changing sea level and a variety of human activities. The water is shallower as waves approach the beach. This causes the wave to slow down and rise up forming a crest that eventually collapses onto the beach.

The persistent movement of water parallel to the beach is referred to as a *longshore current*. This is an important process on beaches composed of sand or finer unconsolidated sediments. These currents move sediment along the shoreline creating a variety of depositional features. Local erosion provides sand or other sediments for beaches, but most of the sand comes from rivers. Almost all of the major rivers in the U.S. have been dammed. The sediment normally transported downstream is deposited in reservoirs behind the dams. Therefore, less sediment is available in many coastal areas and beaches have become narrower.

1. **Longshore Current and Sediment Deposition** (Reference: **Brigantine Inlet**, New Jersey, Quadrangle)

   The **Little Egg Harbor** stereogram shows the development of a sand **spit** (Figure 26.1). Spits are extensions of beach deposits into open water. In this photograph the current is moving from left to right and the spit is also growing in that direction. A second spit, much smaller can be seen at E.3.

   a. The photograph was taken during 1950. The Brigantine Inlet map shows the same area, but was based on photographs taken during 1989. Describe coastal changes during this time.

   b. Based on the map, what direction is the water moving?

   _____

317

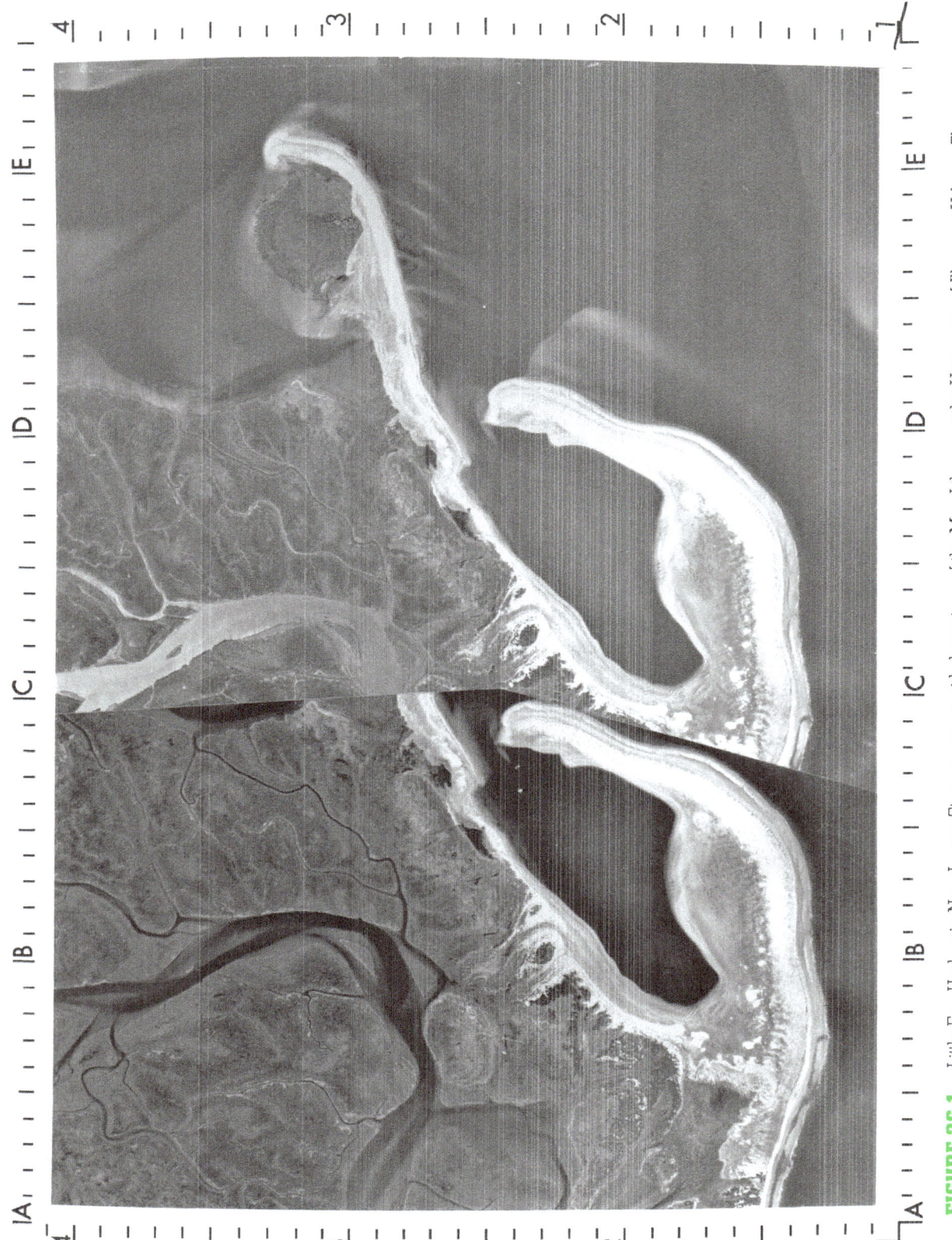

FIGURE 26.1 Little Egg Harbor in New Jersey. Stereogram provided courtesy of the Map Library at the University of Illinois at Urbana-Champaign.

BRIGANTINE INLET, NEW JERSEY

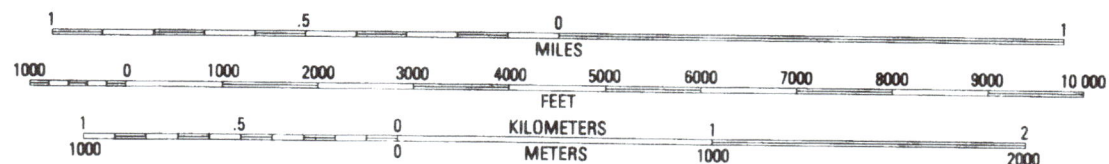

SCALE 1:24,000

Contour Interval 5 Feet

c. What is the contour interval on the map and what is the highest elevation shown?

d. What was the approximate area of the spit during

   1950 (photograph): _____

   1989 (map): _____

e. Is there any evidence of human activity?

2. **Beach Erosion and Use of Jetties** (Reference: **Harwich**, Massachusetts, Quadrangle)

The Harwich Map shows the southern shore of Cape Cod, Massachusetts. The longshore current is from the west toward the east. A series of jetties have been built along the beach extending into the ocean to prevent beach erosion. Figure 26.2 is a stereogram showing the same beach area. The jetties trap sand on the upcurrent side creating a sawtooth appearance of the beach that is easily visible on the photograph.

The meandering stream that drains a reservoir north of West Harwich is a good reference for matching the map and photograph. Long jetties were built at the mouth of this stream to prevent it from becoming blocked with sediment. These are clearly visible on the photograph and shown on the map.

a. The meandering stream and reservoir on the photograph appear black. The ocean however, has a much lighter color, and appears in different shades of gray. What causes the difference in appearance?

b. What is the maximum elevation shown on the map?

   _____

c. What is the maximum water depth shown on the map?

   _____

d. What is the length of the longest jetty extending into the ocean?

   _____

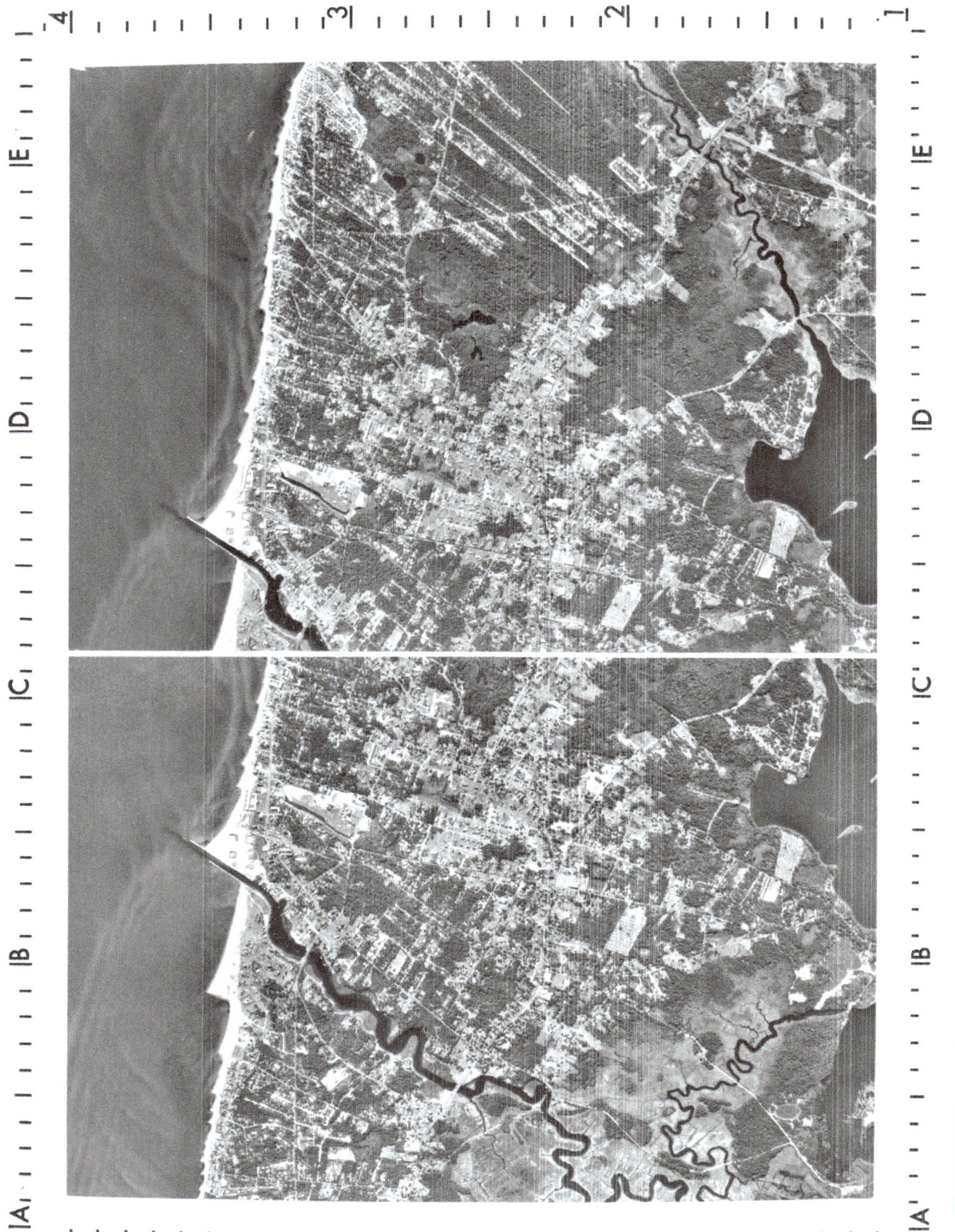

FIGURE 26.2 Cape Cod in Massachusetts. Stereogram provided courtesy of the Map Library at the University of Illinois at Urbana–Champaign.

HARWICH INLET, MASSACHUSETTS

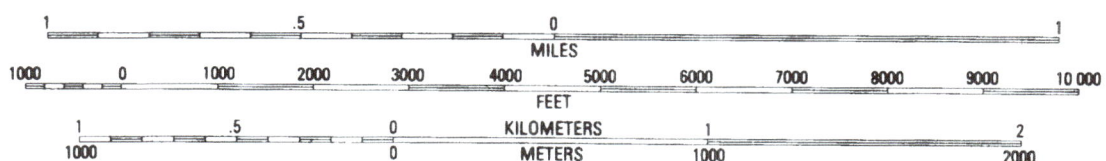

SCALE 1:24,000

Contour Interval 10 Feet

3. **Cliffed Coastlines** (Reference: **Santa Cruz Island A**, California, Quadrangle)

    Cliffed coastlines are notably different from the sandy depositional beaches shown on the two previous maps. **The Santa Cruz Island A** depicts one of the islands off the coast of southern California. These islands have rocky shorelines with steep nearly vertical cliffs. The cliffs shown on this map are highly resistant to erosion, but over long periods of time the rock is worn down by wave action that creates caves and arches. Figure 26.3 is a stereogram showing a part of this map.

    a. What is the height of the cliff at

      Fraser Point? _____

      West Point? _____

    b. What is the highest elevation shown on the map?

      _____

    c. There is a steep sided ridge nearly parallel to the coastline. You can get a good three-dimensional view of the ridge on the stereogram. Describe the stream channels on the ridge. Are these permanent or ephemeral streams? Explain the basis for your answer.

    d. What is the elevation of the ridge crest?

      _____

    e. Describe the vegetation on the island.

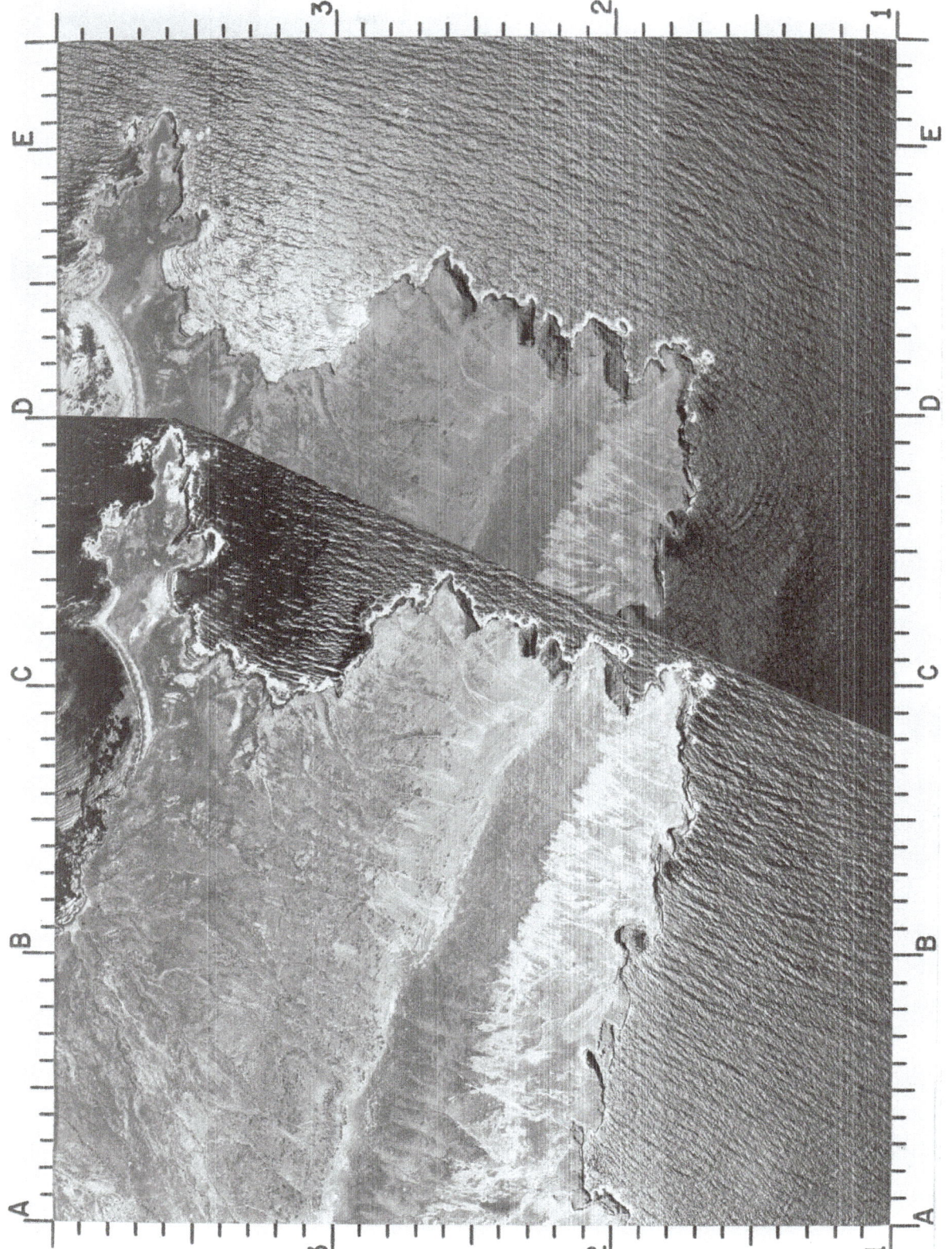

FIGURE 26.3 Fraser Point in California. Stereogram provided courtesy of the Map Library at the University of Illinois at Urbana-Champaign.

SANTA CRUZ ISLAND A, CALIFORNIA

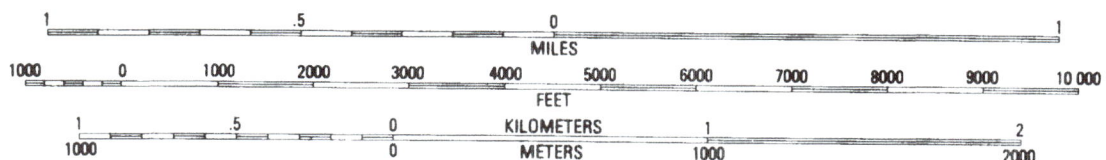

SCALE 1:24,000

Contour Interval 50 Feet

Trigonometry Tables

APPENDIX A

Deg	Sin	Cos	Tan	Cot	Deg
01	.0175	.9998	.0175	57.290	89
02	.0349	.9994	.0349	28.636	88
03	.0523	.9986	.0524	19.081	87
04	.0698	.9976	.0699	14.300	86
05	.0872	.9962	.0875	11.430	85
06	.1045	.9945	.1051	9.514	84
07	.1219	.9925	.1228	8.144	83
08	.1392	.9903	.1405	7.115	82
09	.1564	.9877	.1588	6.313	81
10	.1736	.9848	.1736	5.671	80
11	.1908	.9816	.1944	5.144	79
12	.2079	.9781	.2126	4.704	78
13	.2250	.9744	.2309	4.331	77
14	.2419	.9703	.2493	4.010	76
15	.2588	.9659	.2679	3.732	75
16	.2756	.9613	.2867	3.487	74
17	.2924	.9563	.3057	3.270	73
18	.3090	.9511	.3249	3.077	72
19	.3256	.9455	.3443	2.904	71
20	.3420	.9397	.3640	2.747	70
21	.3584	.9336	.3839	2.605	69
22	.3746	.9272	.4040	2.475	68
23	.3907	.9205	.4245	2.355	67
24	.4067	.9135	.4452	2.246	66
25	.4226	.9063	.4663	2.144	65
26	.4384	.8988	.4877	2.050	64
27	.4540	.8910	.5095	1.962	63
28	.4695	.8829	.5317	1.880	62
29	.4848	.8746	.5543	1.804	61
30	.5000	.8660	.5774	1.732	60

Deg	Sin	Cos	Tan	Cot	Deg
31	.5150	.8572	.6009	1.664	59
32	.5299	.8480	.6249	1.600	58
33	.5446	.8387	.6494	1.539	57
34	.5592	.8290	.6745	1.482	56
35	.5736	.8192	.7002	1.428	55
36	.5878	.8090	.7265	1.376	54
37	.6018	.7986	.7536	1.327	53
38	.6157	.7880	.7813	1.279	52
39	.6293	.7771	.8098	1.234	51
40	.6428	.7660	.8391	1.191	50
41	.6561	.7547	.8693	1.150	49
42	.6691	.7431	.9004	1.110	48
43	.6820	.7314	.9325	1.072	47
44	.6947	.7193	.9657	1.035	46
45	.7071	.7071	1.000	1.000	45

Diameter and Corresponding Circumference and Area

APPENDIX B

Diameter (cm)	Circumference (cm)	Area (m²)
3	9.42	.001
4	12.57	.001
5	15.71	.002
6	18.85	.003
7	21.99	.004
8	25.13	.005
9	28.27	.006
10	31.42	.008
11	34.56	.010
12	37.70	.011
13	40.84	.013
14	43.98	.015
15	47.12	.018
16	50.26	.020
17	53.41	.023
18	56.55	.025
19	56.69	.028
20	62.83	.031
21	65.97	.035
22	69.12	.038
23	72.26	.042
24	75.40	.045
25	78.54	.049
26	81.68	.053
27	84.82	.057
28	87.96	.062
29	91.11	.066
30	94.24	.070
31	97.38	.075

Diameter (cm)	Circumference (cm)	Area (m²)
32	100.53	.080
33	103.67	.088
34	106.81	.090
35	109.95	.096
36	113.09	.101
37	116.23	.107
38	119.38	.113
39	122.52	.119
40	125.66	.125

Index

abandoned meander, 273
acute triangles. *see* Oblique triangle area, trigonometry
adiabatic processes, 87–94
    convection, 89–94
    orographic precipitation, 87–89
adiabatic rate, saturated, 87
aerial photograph interpretation, 148–168
    stereoscopic vision, 163–165
aerial photographs, locating and ordering, 326
air masses, 95–107
albedo of surface, 39
Alice Springs, Australia, 142–143
alluvial fans, Stovepipe Wells, 285–289
altitude, atmospheric pressure and, 53–55
altitude of noon sun, 34
analemma, 37
angular velocity, 27
ANS. *see* Altitude of noon sun
Antarctic Circle, 33
anticyclones, 57
Arctic Circle, 33
arid landscapes, 285–300
Athens, Greece, 146–147
Atlantic basin
    hurricane tracks, 112
    map of, 110
atmospheric humidity, 79–85
atmospheric pressure, 53–61
    winds, 53–61
Autumnal equinox, 29
azimuths, 3

Bangui, Central African Republic, 138–139
bankfull, defined, 238
bankfull discharge, 238–240
baroclinically-enhanced hurricane, 109–111
barometric pressure, 58–60

beach erosion
    Harwich, Massachusetts, 321–324
    jetties, Harwich, Massachusetts, 321–324
Bear Creek, South Dakota, 290–292
Bermuda high, influence on hurricane track, 113–115
braided stream, 281
Brigantine Inlet, New Jersey, 320
    sediment deposition, longshore current, 317–321
    topographic map, 319
bulk density, soil, 207

Cairo, Egypt, 142–143
Cape Cod, Massachusetts, 322
Cape Solitude, Arizona, 268
    topographic map, 267
central meridian, 18
classification of climate, 129–158
clay, 203–204, 233
clay loam, 204
cliffed coastlines, Santa Cruz Island, 325–328
climate classification, 129–158
    dry climates, 141–144
    mid-latitude climates, 145
    procedure, 132–136
    tropical climates, 137–140
coarse sand, 203
coastal geomorphology, 317–328
cold continental climates, 145
Columbus, Christopher, 64
continental glacier, surface erosion, Daley Bay, Minnesota, 309–313
contour mapping, 189–199
    data collection, 189–190
    sample data set, 193–199
convection, 89–94
coordinate determination of distance, 14–15

Coriolis effect, 57
Coriolis force, 63–66
counter-trade winds, 68–70
Crillon Glacier, Alaska, 301–305
cyclones, 57, 95–107, 109–121

Daley Bay, Minnesota, 309–313
    continental glacier, surface erosion, 309–313
DAR. *see* Dry adiabatic rate
day, length of, 30–33
Death Valley, USA, 142–143
declination of sun, 28–29
desert, 141
Dew point, 81
diameter, corresponding circumference, area, 331–332
distance, coordinate determination, 14–15
drumlins, St. Peter, Wisconsin, 313–316
dry adiabatic rate, 87
dry climates, 141–144

earth rotation, 27–28
earth-sun relations, 27–37
    day length, 30–33
    declination of sun, 28–29
    earth rotation, 27–28
    night, length of, 30–33
    noon sun, altitude of, 34–37
eastings, 171
eccentricity code, 45
eccentricity cycle, 46–47
El Niño, 75
ellipse, center of, 47
ELR. *see* Environmental lapse rate
environmental lapse rate, 87
equinoxes, 34
erosion, 217–229, 238–239
    estimation, 221–223, 225
    future soil loss, predicting, 223–224
    gully development, 226
    gully volume calculation, 226–228
    soil, 217–229
    weight of soil loss, 229

Ferrel Cell, 63, 70
field mapping, 1–7
    compass, 1–2
    mapping exercise, 3–7
    measuring azimuth, 1–2
    pacing, 2
    triangulation, 2–3
fine sand, 203

Fishhook Lake, Mississippi, 276
    topographic map, 275
flood frequency curve, recurrence interval, 255
flood recurrence interval, 254–259
fluvial geomorphology, 263–284
focus of ellipse, 47
Fort Mitchell, Alabama-Georgia, 218, 220
    topographic map, 217, 219
Fraser Point, California, 326–327
Fraxinus Pennsylvanian. *see* Green ash
fronts, 95–107
Fuzhou, China, 146–147

geographic grid coordinates, 10
geographic grid, defined, 9
geomorphology
    coastal, 317–328
    fluvial, 263–284
glacial U-shaped valley, Mt. Moran, Wyoming, 305–309
glaciated landscapes, 301–316
global atmosphere, 63–78
    counter-trade winds, 68–70
    ocean currents, 72–74
    polar front jet stream, 70–74
    southern oscillation, 74–78
    surface winds, 72–74
    trade winds, 65–66
    westerlies, 66–68
    wind belts, seasonal shifting of, 68–70
global positioning systems, 169–177
    data collection, 172–177
    Universal Transverse Mercator coordinate system, 169–172
GPS. *see* Global positioning systems
Grand Teton National Park, Wyoming, 306
gravel, 203
great circle, defined, 16
great circle distances, 16–17
gully
    development of, 226
    volume calculation, 226–228

Hadley cell, 63, 70, 75
    surface flow within, 63, 65
Harwich Inlet, Massachusetts, 324
    jetties, beach erosion, 321–324
    topographic map, 323
headward erosion, Scenic, South Dakota, quadrangle, 289–293
Heidelberg, Kentucky, 246
    topographic map, 245

Heyerdahl, Thor, 64–65
Horseshoe Lake, Mississippi, 273–274
humidity, 79–85
Huntsville, Alabama, suburban edge, 232
hurricanes, 109–121
    development, 109–111
    Katrina, 115–116
        coordinates for, 116
        surge heights with, 120
    landfall, 119–121
    movement, 111–115
    storm surge, 119–121
    tracking, 115–117
        Atlantic basin, 112
        Bermuda high influence, 113–115
        map, 117
    tropical-only, 109
hydraulic radius, 241–242

infiltration, 231–236
    water quality issues, 231–236
international date line, 22–23
    defined, 22
Intertropical Convergence Zone, 63, 65, 68, 137
Iquiqeu, Chile, 142–143
Irkutsk, Russia, 146–147
isobars, drawing, 58–61
ITCZ. *see* Intertropical Convergence Zone

jet stream, 71
jetty erosion, Harwich, Massachusetts, 321–324

Katrina hurricane, 115–116
    coordinates for, 116
    surge heights with, 120
Kinshasa, Congo, 138–139
Kon-Tiki, 64–65
Koppen climate classification system, 129, 131–136
    map, 130
Koppen, W., 129

Lake Kabetogama, Minnesota, 310
landscape satellite images, 326
lateral erosion, 269
latitude, 9–25
    distance, 13–14
Liquidambar styraciflua. *see* Sweetgum
Little Egg Harbor, New Jersey, 318
loam, 204
loamy sand, 204
London, England, 146–147

longitude, 9–25
    distance, 13–14
    time, 9–25
longshore current, Brigantine Inlet, New Jersey, 317–321

Manaus, Brazil, 138–139
Marble Canyon, Colorado, 266
March equinox, 29
meandering stream, 273, 289–293
medium sand, 203
Mercator Projection, 12, 20–21
Merqui, Myanmar, 138–139
mid-latitude climates, 145
Milankovitch cycles, 45–51
    eccentricity cycle, 46–47
    obliquity cycle, 48–49
    precessional cycle, 50–51
Milankovitch, Milutin, 45–46, 48
mountain glacier, Mt. Fairweather, Alaska, 301–302
Mt. Fairweather, Alaska, 301–304
    mountain glacier, 301–304
    topographic map, 303
Mt. Moran, Wyoming, 305–309
    glacial U-shaped valley, 305–309
    topographic map, 307
Mt. Saint Helens, 67
    ash fallout distribution, 67
    eruption, 67

Nevens, Nebraska, 281, 284
    topographic map, 283
night length, 30–33
noon sun, altitude of, 34–37
North Platte River, Nebraska, 281–282
North Pole, 13
northings, 170

oblique triangle area, trigonometry, 179–188
obliquity cycle, 45, 48–49
ocean circulation, 63–78
ocean currents, 72–74
orographic precipitation, 87–89

parallels of latitude, 28
particle size, soil, 208
physical properties, soils, 203–204
Pinus taeda. *see* Loblolly pine
plant cover factors, 222–223
Platte City, Missouri-Kansas, 160
    topographic map, 159

polar cell, 63, 70
polar front, 63
polar front jet stream, 70–74
precessional cycle, 45, 50–51
precipitation, 87–94
    convection, 89–94
    curve, 251
    orographic precipitation, 87–89
    recurrence interval, 250–254
precipitation record, 249–250
pressure gradient force, 63
prevailing westerlies, 66
profile descriptions, soil, 202
Pythagorean Theorem, 16

radiation, solar, 39–44
    data collection, 39–41
    intensity of, 41–44
rainfall erosion, 221
    index map, 221
recurrence intervals, 249–250
    flood, 254–259
    precipitation, 250–254
    probability, 250
remnant mountains, Wister, California, quadrangle, 297–300
representation fraction, 150
right triangles, trigonometry, 179–181
Ripley South, Tennessee, 168, 258
    topographic map, 167
roughness coefficient, 242–243
runoff, 231–236
    water quality issues, 231–236

Saffir Simpson scale, 118
Sallie Spring, New Mexico, 293–296
    plateau dissection, 293–297
    topographic map, 295
San Juan River, New Mexico, 294, 297
sand, 203–204
sandy clay, 204
sandy clay loam, 204
sandy loam, 204, 233
Santa Cruz Island, California, 328
    cliffed coastlines, 325–328
    topographic map, 327
SAR. *see* Saturated adiabatic rate
satellite images, 326
saturated adiabatic rate, 87
Scenic
    South Dakota quadrangle, 289–293
        meandering stream, 289–293
    South Dakota, topographic map, 291

Seagrove, North Carolina, 154
    topographic map, 153
sediment deposition, Brigantine Inlet, New Jersey, 317–321
September equinox, 29
Sierra Nevada Mountains, California, 297–298
silt, 203–204
silt-clay loam, 233
silt loam, 204, 233
silty clay, 204
silty-clay loam, 204
slope factor, based on steepness, length, 222
slope gradients, 158–163
soil erodability factor, 221–222
soil erosion, 217–229
    estimation of, 221–223, 225
    future soil loss, predicting, 223–224
    gully development, 226
    gully volume calculation, 226–228
    weight of soil loss, 229
soil horizons, 201
soil properties, 201–215
    bulk density, 207
    data sheets, 209–211
    particle size, 208
    physical properties, 203–204
    profile descriptions, 202
    profiles, 201–202
    surface area, 208
    texture, 204–206
soil size categories, 203
soil textural classes, 204
    infiltration rate, 233
solar day, defined, 27
solar radiation, 39–44
    data collection, 39–41
    intensity of, 41–44
solstices
    illumination of earth, 30
South Pole, 13
South Ripley topographic map, 257
southern oscillation, 74–78
St. Peter, Wisconsin, 313–316
    drumlins, 313–316
    topographic map, 315
standard meridian, 18
standard sea-level pressure, 53
Steppe climate regions, 141
stereoscopic vision, 163–165
Stovepipe Wells, California, 285–286, 288
    alluvial fans, 285–289
    topographic map, 287

stream discharge, 237–247
    bankfull discharge, 239–240
    calculation of, 237–238
    erosion potential, 238–239
    prediction, 244–247
    stream velocity, 240–244
stream gradient, 242
stream incision, 269
stream rejuvenation, 277–278
stream velocity, 240–244
sub-tropical highs, 63
summer solstice, 29, 34
sun, noon altitude, 34–37
surface area, soil, 208
surface weather maps, 95–107
surface winds, 72–74

Taxodium distichum. see Bald cypress
textures of soil, 204–206
The Loop, Utah, 278–280
    topographic map, 279
time, 9–25
topographic contours, 156–163
topographic maps, 148–168
    area calculation, 155–156
    contour lines, 156
    conversion of units, 150–154
    map scale, 150
    slope gradient, 158–163
    topographic contours, 156–163
trade winds, 65–66
trigonometry in geographical field work, 179–188
    field exercise, 182–188
    oblique triangle area, 181–185
    right triangles, 179–181
trigonometry tables, 329–330
tropical climates, 137–140
tropical cyclones, 109–121
    landfall, 119–121
    storm surge, 119–121
tropical-only hurricanes, 109

United States
    map, 61
    weather maps, 95–107
Universal Transverse Mercator coordinate system, 169–172
UTM. *see* Universal Transverse Mercator coordinate system

velocity, stream, 240–244

Walsh Knolls, Utah-Colorado, 272
    topographic map, 271
warm temperate climates, 145
water quality issues, 231–236
weather analysis, 123–127
    data collection, 123–127
weather maps, United States, 95–107
weight of soil loss, 229
westerlies, 66–68
wet-bulb depression, 82–85
White River, Colorado, 270
wind belts, seasonal shifting of, 68–70
winds, 53–61
    atmospheric pressure and, 56
winter solstice, 34
Wister, California, 297–300
    topographic map, 299
world time zones map, 19

Topographic Map Symbols

BOUNDARIES
National
State or territorial
County or equivalent
Civil township or equivalent
Incorporated city or equivalent
Park, reservation, or monument
Small park

LAND SURVEY SYSTEMS
U.S. Public Land Survey System:
 Township or range line
 Location doubtful
 Section line
 Location doubtful
 Found section corner; found closing corner
 Witness corner; meander corner

Other land surveys:
 Township or range line
 Section line
 Land grant or mining claim; monument
 Fence line

ROADS AND RELATED FEATURES
Primary highway
Secondary highway
Light duty road
Unimproved road
Trail
Dual highway
Dual highway with median strip
Road under construction
Underpass; overpass
Bridge
Tunnel

BUILDINGS AND RELATED FEATURES
Building or place of employment: small; large
School; church
Built-up area
Racetrack
Airport
Landing strip
Well (other than water); windmill
Water tank: small; large
Other tank: small; large
Covered reservoir
Gaging station
Landmark object
Campground; picnic area
Cemetery: small; large

RAILROADS AND RELATED FEATURES
Standard gauge single track; station
Standard gauge multiple track
Abandoned
Under construction
Narrow gauge single track
Narrow gauge multiple track
Railroad in street
Juxtaposition
Roundhouse and turntable

TRANSMISSION LINES AND PIPELINES
Power transmission line: pole; tower
Telephone or telegraph line
Aboveground oil or gas pipeline
Underground oil or gas pipeline

CONTOURS
Topographic:
 Intermediate
 Index
 Supplementary
 Depression
 Cut; fill
Bathymetric:
 Intermediate
 Index
 Primary
 Index Primary
 Supplementary

MINES AND CAVES
Quarry or open pit mine
Gravel, sand, clay, or borrow pit
Mine tunnel or cave entrance
Prospect; mine shaft
Mine dump
Tailings

SURFACE FEATURES
Levee
Sand or mud area, dunes, or shifting sand
Intricate surface area
Gravel beach or glacial moraine
Tailings pond

VEGETATION
Woods
Scrub
Orchard
Vineyard
Mangrove

COASTAL FEATURES
Foreshore flat
Rock or coral reef
Rock bare or awash
Group of rocks bare or awash
Exposed wreck
Depth curve; sounding
Breakwater, pier, jetty, or wharf
Seawall

BATHYMETRIC FEATURES
Area exposed at mean low tide; sounding datum
Channel
Offshore oil or gas: well; platform
Sunken rock

RIVERS, LAKES, AND CANALS
Intermittent stream
Intermittent river
Disappearing stream
Perennial stream
Perennial river
Small falls; small rapids
Large falls; large rapids
Masonry dam
Dam with lock
Dam carrying road
Intermittent lake or pond
Dry lake
Narrow wash
Wide wash
Canal, flume, or aqueduct with lock
Elevated aqueduct, flume, or conduit
Aqueduct tunnel
Water well; spring or seep

GLACIERS AND PERMANENT SNOWFIELDS
Contours and limits
Form lines

SUBMERGED AREAS AND BOGS
Marsh or swamp
Submerged marsh or swamp
Wooded marsh or swamp
Submerged wooded marsh or swamp
Rice field
Land subject to inundation

UNITED STATES DEPARTMENT OF THE INTERIOR GEOLOGICAL SURVEY

TOPOGRAPHIC MAP INFORMATION AND SYMBOLS
MARCH 1978

QUADRANGLE MAPS AND SERIES

Quadrangle maps cover four-sided areas bounded by parallels of latitude and meridians of longitude. Quadrangle size is given in minutes or degrees.
Map series are groups of maps that conform to established specifications for size, scale, content, and other elements.
Map scale is the relationship between distance on a map and the corresponding distance on the ground.
Map scale is expressed as a numerical ratio and shown graphically by bar scales marked in feet, miles, and kilometers.

NATIONAL TOPOGRAPHIC MAPS

Series	Scale	1 inch represents	1 centimeter represents	Standard quadrangle size (latitude-longitude)	Quadrangle area (square miles)
7½-minute	1:24,000	2,000 feet	240 meters	7½ × 7½ min.	49 to 70
7½ × 15-minute	1:25,000	about 2,083 feet	250 meters	7½ × 15 min.	98 to 140
Puerto Rico 7½-minute	1:20,000	about 1,667 feet	200 meters	7½ × 7½ min.	71
15-minute	1:62,500	nearly 1 mile	625 meters	15 × 15 min.	197 to 282
Alaska 1:63,360	1:63,360	1 mile	nearly 634 meters	15 × 20 to 36 min.	207 to 281
Intermediate	1:100,000	nearly 1.6 miles	1 kilometer	30 × 60 min.	1568 to 2240
U. S. 1:250,000	1:250,000	nearly 4 miles	2.5 kilometers	1° × 2° or 3°	4,580 to 8,669
U. S. 1:1,000,000	1:1,000,000	nearly 16 miles	10 kilometers	4° × 6°	73,734 to 102,759
Antarctica 1:250,000	1:250,000	nearly 4 miles	2.5 kilometers	1° × 3° to 15°	4,089 to 8,336
Antarctica 1:500,000	1:500,000	nearly 8 miles	5 kilometers	2° × 7½°	28,174 to 30,462

CONTOUR LINES SHOW LAND SHAPES AND ELEVATION

The shape of the land, portrayed by contours, is the distinctive characteristic of topographic maps.
Contours are imaginary lines following the ground surface at a constant elevation above or below sea level.
Contour interval is the elevation difference represented by adjacent contour lines on maps.
Contour intervals depend on ground slope and map scale. Small contour intervals are used for flat areas; larger intervals are used for mountainous terrain.
Supplementary dotted contours, at less than the regular interval, are used in selected flat areas.
Index contours are heavier than others and most have elevation figures.
Relief shading, an overprint giving a three-dimensional impression, is used on selected maps.
Orthophotomaps, which depict terrain and other map features by color-enhanced photographic images, are available for selected areas.

COLORS DISTINGUISH KINDS OF MAP FEATURES

Black is used for manmade or cultural features, such as roads, buildings, names, and boundaries.
Blue is used for water or hydrographic features, such as lakes, rivers, canals, glaciers, and swamps.
Brown is used for relief or hypsographic features—land shapes portrayed by contour lines.
Green is used for woodland cover, with patterns to show scrub, vineyards, or orchards.
Red emphasizes important roads and is used to show public land subdivision lines, land grants, and fence and field lines.
Red tint indicates urban areas, in which only landmark buildings are shown.
Purple is used to show office revision from aerial photographs. The changes are not field checked.

INDEXES SHOW PUBLISHED TOPOGRAPHIC MAPS

Indexes for each State, Puerto Rico and the Virgin Islands of the United States, Guam, American Samoa, and Antarctica show available published maps. Index maps show quadrangle location, name, and survey date. Listed also are special maps and sheets, with prices, map dealers, Federal distribution centers, and map reference libraries, and instructions for ordering maps. Indexes and a booklet describing topographic maps are available free on request.

HOW MAPS CAN BE OBTAINED

Mail orders for maps of areas east of the Mississippi River, including Minnesota, Puerto Rico, the Virgin Islands of the United States, and Antarctica should be addressed to the Branch of Distribution, U. S. Geological Survey, 1200 South Eads Street, Arlington, Virginia 22202. Maps of areas west of the Mississippi River, including Alaska, Hawaii, Louisiana, American Samoa, and Guam should be ordered from the Branch of Distribution, U. S. Geological Survey, Box 25286, Federal Center, Denver, Colorado 80225. A single order combining both eastern and western maps may be placed with either office. Residents of Alaska may order Alaska maps or an index for Alaska from the Distribution Section, U. S. Geological Survey, Federal Building-Box 12, 101 Twelfth Avenue, Fairbanks, Alaska 99701. Order by map name, State, and series. On an order amounting to $300 or more at the list price, a 30-percent discount is allowed. No other discount is applicable. Prepayment is required and must accompany each order. Payment may be made by money order or check payable to the U. S. Geological Survey. Your ZIP code is required.

Sales counters are maintained in the following U. S. Geological Survey offices, where maps of the area may be purchased in person: 1200 South Eads Street, Arlington, Va.; Room 1028, General Services Administration Building, 19th & F Streets NW, Washington, D. C.; 1400 Independence Road, Rolla, Mo.; 345 Middlefield Road, Menlo Park, Calif.; Room 7638, Federal Building, 300 North Los Angeles Street, Los Angeles, Calif.; Room 504, Custom House, 555 Battery Street, San Francisco, Calif.; Building 41, Federal Center, Denver, Colo.; Room 1012, Federal Building, 1961 Stout Street, Denver Colo.; Room 1C45, Federal Building, 1100 Commerce Street, Dallas, Texas; Room 8105, Federal Building, 125 South State Street, Salt Lake City, Utah; Room 1C402, National Center, 12201 Sunrise Valley Drive, Reston, Va.; Room 678, U. S. Court House, West 920 Riverside Avenue, Spokane, Wash.; Room 108, Skyline Building, 508 Second Avenue, Anchorage, Alaska; and Federal Building, 101 Twelfth Avenue, Fairbanks, Alaska.

Commercial dealers sell U. S. Geological Survey maps at their own prices. Names and addresses of dealers are listed in each State index.

INTERIOR—GEOLOGICAL SURVEY RESTON VIRGINIA—1978